PROJECT AIR FORCE

T0290510

Evaluating an Operator Physical Fitness Test Prototype for Tactical Air Control Party and Air Liaison Officers

A Preliminary Analysis of Test Implementation

Sean Robson, Tracy C. McCausland, Jennifer L. Cerully, Stephanie Pezard, Laura Raaen, Nahom M. Beyene

Prepared for the United States Air Force
Approved for public release; distribution unlimited

For more information on this publication, visit www.rand.org/t/RR2171

Library of Congress Cataloging-in-Publication Data is available for this publication.
ISBN: 978-0-8330-9944-0

Published by the RAND Corporation, Santa Monica, Calif.
© Copyright 2018 RAND Corporation
RAND® is a registered trademark.

Support RAND
Make a tax-deductible charitable contribution at
www.rand.org/giving/contribute

www.rand.org

Preface

In January 2013, the Chairman of the Joint Chiefs of Staff and the Secretary of Defense rescinded the 1994 Direct Ground Combat Definition and Assignment Rule and mandated that "[v]alidated gender-neutral occupational standards will be used to assess and assign Service members not later than September 2015" (Chairman of the Joint Chiefs of Staff, 2013, p. 2). In support of this mandate, the U.S. Air Force asked the RAND Corporation to assist its development and validation of gender-neutral tests and standards for battlefield airmen (BA) specialties, which, at the time, were the only occupational specialties that remained closed to women in the Air Force. The Air Force complied with this mandate by conducting an extensive validation study, which examined the occupational relevance of physical fitness (PF) tests and standards. Following the fitness test validation study, one enlisted specialty—Tactical Air Control Party—and one officer BA specialty—Air Liaison Officer—moved forward with an implementation plan to further evaluate a set of recommended tests and continuation standards. This report describes RAND's assistance to the Air Force on two fronts: (1) conducting a preliminary evaluation of potential issues and concerns that might influence implementation effectiveness and (2) developing a framework for evaluating the implementation of occupationally relevant and specific tests and standards. This work will provide the foundation for ongoing review and evaluation of Air Force fitness tests and standards, which are designed to ensure airmen are capable of performing critical physical tasks associated with their assigned specialties.

This report describes research that should be relevant to policy and research audiences with interests in PF standards and the implementation of physical tests and standards for military occupations. The work in this report was cosponsored by the Air Force Director of Military Force Management Policy, Deputy Chief of Staff for Manpower, Personnel and Services, the Vice Commander in Air Education and Training Command, the Vice Commander in Air Force Special Operations Command, and the Directorate of Air and Space Operations. This research was conducted within the Manpower, Personnel, and Training Program of RAND Project AIR FORCE.

RAND Project AIR FORCE

RAND Project AIR FORCE (PAF), a division of the RAND Corporation, is the U.S. Air Force's federally funded research and development center for studies and analyses. PAF provides the Air Force with independent analyses of policy alternatives affecting the development, employment, combat readiness, and support of current and future air, space, and cyber forces. Research is conducted in four programs: Force Modernization and Employment;

Manpower, Personnel, and Training; Resource Management; and Strategy and Doctrine. The research reported here was prepared under contract FA7014-16-D-1000.

Additional information about PAF is available on our website: www.rand.org/paf.

This report documents work originally shared with the U.S. Air Force on September 30, 2016. The draft report, issued on September 30, 2016, was reviewed by formal peer reviewers and U.S. Air Force subject-matter experts.

Contents

Figures

Tables

Summary

In January 2013, then–Chairman of the Joint Chiefs of Staff Martin Dempsey and then–Secretary of Defense Leon Panetta issued a memorandum rescinding the 1994 Direct Ground Combat Definition and Assignment Rule, which excluded women from assignment to units and positions whose primary mission is to engage in direct combat on the ground. In the memorandum, Panetta and Dempsey mandated that "[c]urrently closed units and positions will be opened by each relevant Service . . . after the development and implementation of validated, gender-neutral occupational standards and the required notification to Congress" (Chairman of the Joint Chiefs of Staff, 2013). To comply with this mandate, the U.S. Air Force, with the assistance of the RAND Corporation, established a process in fiscal year (FY) 2012 to identify and validate gender-neutral tests, standards, and physical requirements. This initial effort was followed with additional technical support by RAND in FY 2014 and 2015 for the Air Force's implementation of an extensive criterion-related validation study based on scientific principles. This study led to the recommendation of a ten-component Operator–Prototype Test Battery (O-PTB) for the enlisted Tactical Air Control Party (TACP) specialty and the Air Liaison Officer (ALO) specialty.[1] A list of these tests is provided in Table S.1 and a full description and protocol for each test is provided in Appendix A.

Table S.1. Physical Fitness Tests in the O-PTB

1. Grip Strength	6. Lunges, 50-lb Sandbag
2. Medicine Ball Toss (Backwards, Sidearm, Overhead)	7. Extended Cross Knee Crunch
3. Three-Cone Drill	8. Farmer's Carry
4. Rhomboid Major (RM) Trap Bar Deadlift	9. Ergometer Row Test (1,000 meters)
5. Pull-Up Test	10. Run (1.5 miles)[a]

[a] The 1.5-mile run test was not administered as part of the Air Force Exercise Science Unit's (AF-ESU's) implementation trips because the Air Force already conducts this test regularly as part of the Tier I Air Force–wide fitness test.

---

[1] Because recruits and trainees already have a set of tests and standards in place for TACP and ALO, the Air Force prioritized the implementation of tests and standards for existing TACP and ALO operators. Future plans may consider updating the tests and standards used for recruits and trainees as well as other battlefield airmen (BA) specialties.

The AF-ESU is now leading efforts to implement these tests. As part of this effort, the AF-ESU developed an implementation, verification, and training (IVT) plan to address several questions such as:

- Who will administer the tests?
- How long will it take to administer and take the tests?
- What is the likelihood of a test-taker sustaining an injury while taking the tests?
- How many current TACPs and ALOs would be expected to pass the proposed test standards?
- How much improvement can be expected in test performance as TACPs and ALOs become more familiar with the tests?
- How well does performance on the test battery differentiate between successful and less successful TACPs and ALOs?
- What concerns do the different stakeholders have about the tests and standards?

Purpose and Approach of the Research

Even though the AF-ESU prioritized these questions, other short- and long-term issues and concerns should be identified and may need to be subsequently addressed. Consequently, RAND was asked by Air Force Director of Military Force Management Policy, Deputy Chief of Staff for Manpower, Personnel and Services to offer support by designing a preliminary evaluation effort of the implementation of the physical tests and standards being adopted by the TACP and ALO career fields. Our evaluation emphasizes one of the main IVT questions: the concerns of different stakeholders. After reviewing initial results from this evaluation, we developed a broad evaluation framework to identify other possible issues and concerns that may emerge in relation to the implementation of physical tests and standards.

Our approach to achieve these two objectives consisted of four steps:

1. Identify relevant stakeholders.
2. Design evaluation instruments to address stakeholder reactions.
3. Collect and analyze the data.
4. Develop an evaluation framework for future evaluation efforts.

Step One: Identify Relevant Stakeholders

The Air Force plans to sequence the tests and standards implementation in three waves that correspond to three primary job roles for TACPs and ALOs. Specifically, as tests and standards are adopted, they will first be implemented for TACP and ALO operators, then technical training students, and finally for recruits. At this time, the Air Force is in its first wave of implementation (for TACP and ALO operators). Therefore, the most immediate priority for our preliminary evaluation efforts is to determine stakeholder groups involved in these initial implementation efforts for TACP and ALO operators. Three primary stakeholder groups were identified:

- TACP and ALO operators, who will be required to take the tests and meet the specified standards
- physical training leaders (PTLs), who will be responsible for administering and scoring the tests and providing training to other test administrators
- career field managers (CFMs), who will be responsible for addressing gaps in readiness levels for their specialty and for overseeing whether resource needs are being met across the career field.

Step Two: Design Evaluation Instruments to Address Stakeholder Reactions

Next, we identified several topics that could affect the successful implementation of the O-PTB. Taking into consideration the specific topics most relevant to each stakeholder group and prior research on test-taker reactions, we identified the following primary topics for each stakeholder group:

- operator perspective (TACPs and ALOs)
  - consistency of test administration
  - knowledge of test performance relative to test standards
  - injury concerns related to the tests
  - experienced levels of frustration in taking the tests
  - global evaluations of the O-PTB, including perceived utility, validity, and fairness

- PTL perspective
  - quality of training provided to administer and score tests
  - global evaluations of the O-PTB, including perceived utility, validity, and fairness

- CFM perspective
  - current and future plans for test implementation
  - perceived benefits of implementing a new test battery
  - perceived challenges or drawbacks of implementing a new test battery
  - concerns with the specific recommended test battery
  - specific barriers and concerns for test implementation such as time to administer, potential for injury, cost, fairness, and utility.

To address these topics, we used a mixed-methods approach. Specifically, we developed evaluation surveys for TACP and ALO operators and for PTLs and conducted semistructured interviews with CFMs.

Step Three: Collect and Analyze the Data

Evaluation surveys were administered to TACPs, ALOs, and PTLs during the AF-ESU implementation trips to different installations. For each implementation trip, the AF-ESU trained the PTLs on how to properly administer and score each test in the O-PTB and then administered tests to a sample of TACPs and ALOs available at the time of the trip.

Evaluation surveys were completed by a total of 198 operators (TACPs and ALOs) and 135 PTLs representing units from 12 different installations. We also conducted semistructured interviews with CFMs both from the TACP and ALO specialties, as well as other BA specialties including Pararescue (PJ), Combat Rescue Officer (CRO), Special Tactics Officer (STO), and Special Operations Weather (SOWT).

Step Four: Develop an Evaluation Framework for Future Evaluation Efforts

Our initial evaluation is focused on topics relevant to immediate implementation priorities of the Air Force, however, we recognize a broader perspective on evaluation is beneficial to identify potential future concerns or issues that emerge as priorities over time (e.g., focus on tests and standards for recruits). Consequently, we present a more comprehensive framework identifying a range of possible topics that are organized in a framework often used in the military for identifying requirements and potential gaps for a given set of strategic objectives.[2] Broadly, the objectives of the evaluation framework are to (1) raise awareness of potential challenges and concerns for relevant stakeholders during the implementation and adoption of the new physical tests and standards and (2) promote the development of systematic data collection to monitor progress over time.

Results of Evaluation Surveys and Semistructured Interviews

In the following sections, we present an overview of results from each of the stakeholder groups. We begin by discussing results from the evaluation surveys from the TACP and ALO operators and then PTLs. We conclude with a summary of themes identified during the discussions held with CFMs.

TACP and ALO Survey Responses

Consistency of Test Administration

The pattern of TACP and ALO responses suggested that each test was administered consistently. Specifically, over 90 percent of respondents agreed or strongly agreed that each test was consistently administered. Furthermore, only two tests, Pull-Ups and Extended Cross Knee Crunch, had less than 95 percent agreeing or strongly agreeing that these tests were administered consistently (94 and 92 percent, respectively).

Knowledge of Performance Relative to Standard

TACP and ALOs indicated that they knew how well they performed on each test relative to the required standard established as part of the Air Force validation study. Only the Extended

---

[2] Doctrine, Organization, Training, Materiel, Leadership and Education, Personnel, and Facilities and Policy (DOTMLPF-P) is defined in the Joint Capabilities Integration Development System.

Cross Knee Crunch yielded less than 90 percent of responses agreeing or strongly agreeing, with 9 percent of respondents indicating they neither agreed nor disagreed with the statement.

Injury Concerns for Each Test

Overall, TACP and ALO respondents noted few injury concerns across the majority of the tests, and no new injuries were reported in preparation for or during administration of the tests. However, 20 percent of respondents indicated an injury concern for the Trap Bar Deadlift. Additional analyses indicated no patterns in responses for the Trap Bar Deadlift across subgroups. However, respondents' open-ended comments suggest concerns primarily emphasized the ability to use proper form and technique when performing the lift. Some of the respondents provided specific suggestions to address potential injury concerns, including providing sufficient training on proper form, providing opportunities to practice, and using a weight belt when lifting.

Frustration Experienced

Tests that produced the most frustration included the Extended Cross Knee Crunch (41 percent) and the Medicine Ball Toss (21 percent); these tests demonstrated significantly more frustration compared with other tests in the test battery. We conducted additional analyses for these two tests to determine if any significant patterns emerged by installation and background characteristics.

Although installation was not influential, comparisons of responses by demographic variables revealed that TACPs and ALOs who are heavier, taller, and more experienced were significantly less likely to indicate frustration on the Medicine Ball Toss compared with lighter, shorter, and less experienced TACPs and ALOs. This finding is further supported by a few open-ended comments suggesting that the Medicine Ball Toss appears to favor taller individuals. Other common open-ended comments indicated that some respondents felt that the Medicine Ball Toss required substantial technique and practice to do well, with some stating that the test measures skill/technique more than strength/power.

Global Evaluations of the Test Battery

Several items were used to address operators' perspectives on the perceived utility of the test battery, in addition to perceptions of how well they think the test battery measures important abilities for their job (i.e., face validity). Overall, the general perceptions suggest that the test battery will be positively received by TACPs and ALOs, with 84 percent of respondents indicating that knowing how well they performed on this test battery will help improve job-related physical capabilities; 74 percent indicated that this test measures the abilities required of a TACP.

Although most responses present favorable perceptions, 18 percent indicated that they disagreed or strongly disagreed that the test would be fair for all TACPs regardless of rank, age, stature, gender, or race/ethnicity. Respondents' open-ended comments suggested that the tests and standards might not be relevant to all TACPs given that some assignments do not require any (or as much) physical effort to perform their job tasks.

PTL Survey Responses

PTL Training

We also developed survey items for the PTLs who received test administrator training from the AF-ESU on how to properly administer and score each test. Overall, the PTL responses to the training clearly suggest that the training provided was effective.

Global Evaluations of the Test Battery

PTLs also provided responses to general questions about the prototype test battery. Overall, these responses were also favorable. Ninety-three percent of respondents agreed or strongly agreed that each test was administered to all operators in the same way. Eighty percent of PTLs indicated that the test battery would be fair to all TACPs regardless of rank, age, stature, gender, or race/ethnicity. Some of those who indicated concerns about fairness provided open-ended comments suggesting that fitness requirements may not be relevant across the entire career lifecycle of a TACP. That is, TACPs may be assigned to a position in which fitness is less important, such as a staff position. Other open-ended comments suggest that some of the frustration observed in taking specific tests (e.g., Extended Cross Knee Crunch) may be related to lack of familiarity in taking a new test and a desire to perform well.

Perceived Utility of the Test Battery

The final set of survey items for PTLs focused on the overall utility of the test battery. Responses indicated very positive support with 97 percent agreeing or strongly agreeing that the test battery is a better measure of operational capabilities compared with the Physical Ability and Stamina Test (PAST).[3] Furthermore, 94 percent agreed or strongly agreed that operators could really show their physical abilities through this test battery. And, 92 percent indicated that TACPs would find their test results useful for improving their job-related physical capabilities.

CFM Discussions

To gain a better understanding of the BA CFMs' concerns about test implementation, we held three separate meetings with individual and small groups of CFMs, which included CFMs and career field representatives from TACP and ALO; CRO, PJ, and STO; and SOWT. One

---

[3] We use the PAST as a point of comparison because many of its test components (e.g., pull-ups, push-ups, sit-ups, timed runs) are commonly used across the BA specialties at different career stages to include recruits, trainees, and operators. At a given career stage and/or for each BA specialty, however, there are variations in the standards (e.g., time, distances), ordering of test administration, and/or specific tests (e.g., inclusion of swimming tests).

CFM (Combat Control) was unavailable for these meetings. Although only the TACP and ALO specialties have moved forward with implementation of a new operator test battery, we decided to include other BA CFMs to provide a more comprehensive understanding of the issues, concerns, and barriers influencing the successful implementation of an updated physical test battery. Given the differences in current and planned efforts to implement an updated physical test battery, we present our discussion of the CFM feedback for each topic area grouped by TACP/ALO CFMs and other BA CFMs. Several questions were presented to address the following topics:

1. current and future plans for test implementation
2. perceived benefits of implementing a new test battery
3. perceived challenges or drawbacks of implementing a new test battery
4. concerns with the specific recommended test battery
5. specific barriers and concerns for test implementation such as time to administer, potential for injury, cost, fairness, and utility.

Overall, the CFMs agreed that there are potential benefits gained from the proposed O-PTB; however, other BA CFMs raised concerns about the need for ten tests, the time to administer them, and the potential costs associated with purchasing and maintaining the tests (see Table S.2).

Table S.2. Summary of CFMs' Perspectives on the Implementation of Tests and Standards

Topic	TACP/ALO CFM	Other BA CFM
Plans for test implementation	• Focus is on implementation for current operators • Future efforts will consider implementation for technical training students and future recruits	• General interest and will monitor how well tests and standards are implemented for TACP/ALO • No concrete plans to change PF tests and standards
Perceived benefits	New O-PTB is more comprehensive and addresses deficiencies in the PAST by measuring job-related agility	New O-PTB addresses deficiency in the PAST by measuring muscular power
Perceived challenges	• Tests and standards may not be equally relevant for TACPs/ALOs at all organizational levels (e.g., staff position) • Fear of the unknown (e.g., potential career consequences if operators do not meet standards)	• Administration time is perceived to take more time and will be more difficult to manage compared with the test battery they currently use for their specialties • Additional equipment, which will require more money to purchase and maintain
Concerns with recommended test battery	Concern that commanders may overemphasize or underemphasize role of fitness for specialties; emphasized importance of balanced integration of fitness-related workouts and testing	• Insufficient communication on how final ten PF tests were selected from 39 considered in the validation study • Questioned value added with proposed tests and standards
Other comments	Communication often gets lost over time and not all operators recall steps taken to develop recommended tests and standards	Recognize the value in making improvements, but do not agree that O-PTB is the correct solution

Conclusions

We conducted a preliminary evaluation of potential issues and concerns that may influence implementation effectiveness for TACP and ALO operators and developed a broader framework for monitoring and evaluating the implementation of an occupationally specific PF test battery for TACPs and ALOs. A summary of the main findings and recommendations is provided in Table S.3.

Table S.3. Summary of Findings and Recommendations

Finding	Accompanying Recommendation
1. Overall, TACPs and ALOs indicated consistently strong, positive support for the O-PTB.	Communicate results broadly throughout the TACP and ALO community; disseminate results to other BA leaders.
2. TACPs and ALOs generally felt that each test was administered to all operators in the same way and that they knew how they performed relative to the standard.	Test administrators were being observed and had just received training on how to administer; therefore, follow-up evaluations should be conducted to ensure consistent administration continues to be followed.
3. TACPs and ALOs indicated concern about the potential for injury for the Trap Bar Deadlift (20 percent), and PTLs expressed concern that operators could be injured while taking the test battery (12 percent). No new injuries reported in preparation for or during administration of the tests.	Consider (1) further training on proper form and technique, (2) increasing the opportunities to practice the tests and receive feedback, and/or (3) modifying test administration instructions.
4. TACPs and ALOs were most frustrated by the Extended Cross Knee Crunch (41 percent) and the Medicine Ball Toss (21 percent), and 15 percent of PTLs indicated that operators seemed frustrated by the test.	
5. TACPs and ALOs (18 percent) and PTLs (12 percent) did not feel that the test battery would be fair for all TACPs regardless of rank, age, stature, gender, or race/ethnicity. TACP and ALO CFMs echoed this sentiment by expressing concern about the community's lack of awareness regarding the scientific validation process to determine the tests included in the battery.	Deliver additional communication about the history of the test development process, how tests were selected, and how they link to job and mission-related requirements.
6. TACPs, ALOs, (71 percent) and PTLs (78 percent) felt that it was important that test administrators be other TACPs; in contrast, CFMs emphasized that the test administrator could be anyone.	Consider the advantages and disadvantages of various test administrator characteristics.
7. Other BA CFMs (i.e., not TACP or ALO) expressed strong concerns over logistical issues regarding test administration (e.g., time, equipment cost).	Examine the time required, on average, to administer the prototype test battery and the cost to purchase all equipment for a squadron of 100.
8. Other BA CFMs (i.e., not TACP or ALO) recognized deficiencies in the PAST and expressed interest in addressing these shortcomings through a more collaborative effort.	Increase frequency of communication among CFMs, commanders, strength and conditioning coaches, and the AF-ESU. Consider trade-offs between scientific validity and other career field needs including feasibility, cost, and perceived utility.

Overall, the results from TACPs, ALOs, and PTLs were very positive, with relatively few concerns identified among a minority of TACP and ALO operators who participated in this phase of test implementation. Most indicated that each test was administered correctly and that the test battery measured important job-related physical abilities. Many open-ended comments further supported these findings with positive comparisons to the PAST, which stated that the O-PTB is more comprehensive and more representative of operational tasks than the PAST. Only a few tests caused frustration or raised concerns for potential injury among some of the respondents. Three tests, in particular, warrant close monitoring and further review to identify opportunities to address potential concerns for injury and frustration: the Extended Cross Knee Crunch, Trap Bar Deadlift, and Medicine Ball Toss. For each of these tests, we recommend additional training, practice, and feedback on the proper form and technique, which should help address these concerns.

Analysis of open-ended comments in combination with CFM feedback suggests that additional structured communication could help address a range of concerns (e.g., career repercussions if operators fail a test) and to further educate the operators on the purpose of the test battery and the steps that were taken to inform the decision to select the final ten tests and standards. Consequently, we recommend developing additional communication channels such as short, informative pamphlets that can be used to answer frequently asked questions. For example, operators could benefit from additional education on the role of each test in assessing their physical capability to perform critical physical tasks of their specialty. Operators could also benefit by having additional information on the implementation plan, timeline, and potential changes that may be made to the protocols for specific tests (e.g., Farmer's Carry).

Although the results were very positive, our analysis represents only a snapshot of perceptions at a given point in time. As the implementation proceeds for TACP and ALO operators, we recommend following up with future evaluations to determine if perceptions have changed. Future evaluations should be conducted annually for the first two years of implementation and then every three to five years thereafter. Tracking individual operators over time using pre- and post-test research designs will allow for more sophisticated analyses for evaluating the effectiveness of different interventions (e.g., additional training on form/technique) to address concerns. Additional interviews, focus groups, and surveys of stakeholders can also help to determine whether new issues or concerns have emerged.

Finally, we recommend expanding these evaluation efforts to consider additional stakeholder groups and topics. To assist in identifying other stakeholder groups and topics, we developed a framework guided by the DOTMLPF-P structure. This framework could be used to raise awareness of possible issues that may influence the successful implementation of PF tests and standards and to guide evaluation efforts for determining how well implementation objectives are being met.

Acknowledgments

The design, development, and execution of this study, which adhered to professional and scientific guidelines, was possible only through a great deal of coordination and support from Dr. Neal Baumgartner, the study's point of contact and chief of the Air Force Exercise Science Unit. Baumgartner and his team supported the preliminary evaluation through data collection and related feedback collected during site visits.

Brig Gen Brian Kelly, Air Force Director of Military Force Management Policy, Deputy Chief of Staff for Manpower, Personnel and Services, was the project sponsor during the course of this research. General Kelly adopted the recommendations and methodology to validate gender-neutral occupational standards for battlefield airmen (BA) based on RAND's prior work in fiscal years 2012 and 2014–2015. His continued support provided us with the opportunity to develop a comprehensive framework for evaluating the implementation of occupationally relevant, occupationally specific physical tests and standards.

We would like to thank all of the enlisted Tactical Air Control Party specialty and Air Liaison Officer specialty members who completed the evaluation feedback form and the BA career field managers and subject-matter experts, including CMSgt Michael Bender (ret.), Maj Michael Hayek, Maj Justin Bañez, SMSgt Kenneth Lee Blakeney, Steven Buhrow, Lt Col Charles Bris-Bois, CMSgt Thomas Rich, Lt Col Travis Woodworth, and CMSgt Ronald Richards, who participated in the workshop and small group discussions.

Finally, we would like to recognize David Schulker and Christopher Maerzluft for their technical support on data analyses and Gabriella Gonzalez, Laura Miller, and Agnes Schaefer for their feedback on the evaluation framework.

Abbreviations

AF-ESU	Air Force Exercise Science Unit
AFSC	Air Force Specialty Code
ALO	Air Liaison Officer
BA	battlefield airmen
BPM	beats per minute
CCT	Combat Control
CFM	career field manager
CPT	critical physical task
CRO	Combat Rescue Officer
DOTMLPF-P	Doctrine, Organization, Training, Materiel, Leadership and Education, Personnel, and Facilities and Policy
E	explanations
FC	formal characteristics
FY	fiscal year
IT	interpersonal treatment
IVT	implementation, verification, and training
JBLM	Joint Base Lewis-McChord
O-PTB	Operator–Prototype Test Battery
PF	physical fitness
PJ	Pararescue
PAST	Physical Ability and Stamina Test
PTL	physical training leaders
PTS	physical task simulation
RPE	rate of perceived exertion
SME	subject-matter expert
So	social

SOWT	Special Operations Weather Team
St	structure
STO	Special Tactics Officer
TACP	Tactical Air Control Party

1. Introduction

In January 2013, then–Chairman of the Joint Chiefs of Staff Martin Dempsey and then–Secretary of Defense Leon Panetta issued a memorandum rescinding the 1994 Direct Ground Combat Definition and Assignment Rule, which excluded women from assignment to units and positions whose primary mission is to engage in direct combat on the ground. In the memorandum, Panetta and Dempsey mandated that "[c]urrently closed units and positions will be opened by each relevant Service . . . after the development and implementation of validated, gender-neutral occupational standards and the required notification to Congress" (Chairman of the Joint Chiefs of Staff, 2013). To comply with this mandate, the U.S. Air Force, with assistance from the RAND Corporation, established a process in fiscal year (FY) 2012 to identify and validate gender-neutral tests, standards, and physical requirements. This initial effort was followed with additional technical support by RAND in FYs 2014 and 2015 for the Air Force's implementation of an extensive criterion-related validation study based on scientific principles. Specifically, the Air Force Exercise Science Unit (AF-ESU) conducted a physical task and demands analysis to identify critical physical tasks (CPTs) for several Air Force occupational specialties, including Tactical Air Control Party (TACP), which is currently the only battlefield airmen (BA) specialty that does not have occupationally relevant,[1] occupationally specific physical fitness (PF) continuation standards (e.g., annual fitness testing).

The Air Force and RAND used these CPTs as the foundation for developing physical task simulations (PTSs) to serve as the primary outcome performance measures in the validation study. Each PTS was specifically designed to approximate occupationally relevant CPTs. The PTSs were reviewed by subject-matter experts (SMEs) from target occupational specialties and were evaluated on several criteria (e.g., accuracy of distance, weight) during a pilot test before finalization for the validation study.

The CPTs also served as the basis for AF-ESU's identification of possible PF tests to further evaluate in the validation study. The AF-ESU unit initially identified more than 100 tests for consideration. To narrow the list of tests for the validation study, each test was first scored by a small group of exercise science experts on several criteria (e.g., cost, validity evidence, reliability evidence, potential for injury). This initial review reduced the number of viable tests to 65. These tests were then further evaluated in a pilot test to identify and remove tests that were difficult to administer, received consistently poor feedback from test-takers, or did not correlate well with

---

[1] BA have several specialties, including four enlisted specialties (combat control [CCT], pararescue [PJ], special operations weather [SOWT], and TACP) and three officer specialties (air liaison officer [ALO], combat rescue officer [CRO], and special tactics officer [STO]).

pilot test PTSs. The final list of 39 tests used in the validation study, organized by the PF component,[2] is presented in Table 1.1.

Table 1.1. Physical Fitness Tests Used in the Air Force Validation Study

Power	Agility	Strength	Endurance	Anaerobic	Aerobic
Seated Chest Pass	Three Cone Drill	Grip Strength	Core 1. Side Bridge 2. Cross Knee Crunch 3. Sit-ups	Sled Haul, 70 ft.	3-Mile Ruck March (50-lb. load)
Standing Long Jump	5-10-5 Pro-Agility	Weighted Pull-Ups (25-lb. vest)	Pull-Ups 1. Regular 2. Metronome 3. Alternate	1,000-Meter Row (ergometer)	
			500-Meter Row (ergometer)		
Medicine Ball Toss	Hurdle	Trap Bar Deadlift	Push-ups (metronome)		800-Meter Run
	Y-Balance	Upright Row	Weighted Lunges (50 lbs.)	100-Yard Farmer's Carry	1.5-Mile Run
		Strength Aptitude Test	Heel Touch	300-Yard Shuttle Run	500-Meter Surface Swim
			Overhead Pass		1,500-Meter Fin Swim
			Water Skills 1. 25-Meter Underwater Swim 2. Tread Water 3. Snorkel		Loaded Step-Up (5 minutes)
			Squat (unloaded)		Versa Climber
			Back Suspension		

The validation study included a sample of airmen from a variety of backgrounds, including operators across the BA specialties, and technical training students ($n = 71$), non-BA men ($n = 38$) and non-BA women ($n = 62$). These airmen volunteered to participate in the study by completing a wide range of PF tests and PTSs over a two-week period. Scores of the PF tests were then evaluated to determine how well each test and combination of tests predicted

---

[2] The AF-ESU developed this framework as a guide for the selection of tests and recognizes that tests may be categorized under multiple fitness components.

performance on each PTS. The final set of recommended tests for TACPs and Air Liaison Officers (ALOs) included the tests shown in Table 1.2.[3]

Table 1.2. Tests in the Operator–Prototype Test Battery

1.	Grip Strength	6.	Lunges, 50-lb. sandbag
2.	Medicine Ball Toss (Backwards, Sidearm, Overhead)	7.	Extended Cross Knee Crunch
3.	Three-Cone Drill	8.	Farmer's Carry
4.	RM Trap Bar Deadlift	9.	Ergometer Row Test (1,000 meters)
5.	Pull-Up Test	10.	Run (1.5 miles)[a]

[a] The 1.5-mile run test was not administered as part of the AF-ESU's implementation trips because the Air Force already conducts this test regularly as part of the Tier I Air Force–wide fitness test.

Although these ten tests were generally found to be strong predictors of performance on the task simulations (Robson, forthcoming), several questions remain to be addressed prior to full implementation for the TACP and ALO communities (see Figure 1.1 for more information about these communities).

---

[3] The protocol for each test is presented in Appendix A.

Figure 1.1. TACP and ALO Roles

Tactical Air Control Party

Finds, fixes, tracks, targets, and engages enemy forces in close proximity to friendly forces and assesses strike results. Plans, coordinates, and directs manned and unmanned, lethal and nonlethal air power utilizing advanced command [and] control communications technologies and weapon systems in direct ground combat. Controls and executes air, space, and cyber power across the full spectrum of military operations. Provides airspace deconfliction, artillery, naval gunfire, intelligence, surveillance, and reconnaissance and terminal control of close air support to shape the battlefield. Operates in austere combat environments independent of an established airbase or its perimeter defenses. Employed as part of a joint, interagency, or coalition force, aligned with conventional or special operations combat maneuver units to support Combatant Commander objectives. Primarily assigned to U.S. Army Installations (Air Force Personnel Center, 2016, p. 47).

Air Liaison Officer

The ALO specialty (13LX) integrates Joint Fires during joint and multinational operations. ALOs develop joint fires support plans in the course of the targeting cycle to integrate lethal and nonlethal effects during deliberate and dynamic targeting. They participate in target product development, weaponeering, collateral damage estimation; provide assessment of munitions effectiveness and battle damage; and deliver reattack recommendations. As the direct representatives of the Joint/Multinational Force Air Component Commander, ALOs are the primary Air Force advisers to U.S. Army, joint, multinational, and special operations ground force commanders for the integration of air, space, and cyber power. ALOs plan, request, coordinate, and control close air support as a Joint Terminal Attack Controller. They synchronize and integrate combat airspace; artillery and naval gunfire; and intelligence, surveillance, and reconnaissance. They also assign aircraft to ground force immediate requests for air support. ALOs provide command and control of air-ground operations within their assigned ground force operations area. Furthermore, they lead, plan, organize, and supervise day-to-day TACP Weapons System activities in-garrison and forward deployed. As battlefield airmen, ALOs operate under the most austere conditions for extended periods, independent of an established airbase or its perimeter defenses. When deployed with tactical ground forces, ALOs employ small unit tactics, conduct close quarters battle, casualty collection, vehicle operations, and prepare deployed sites (Air Force Personnel Center, 2017, p. 56).

Recognizing the need for further evaluation, the AF-ESU developed an implementation, verification, and training (IVT) plan to address several questions:

- Who will administer the tests?
- How long will it take to administer and take the tests?
- What is the likelihood of a test-taker sustaining an injury while taking the tests?
- How many current TACPs and ALOs would be expected to pass the proposed test standards?
- How much improvement can be expected in test performance as TACPs and ALOs become more familiar with the tests?
- How well does performance on the test battery differentiate between successful and less successful TACP and ALO occupationally relevant physical task performance?
- What concerns do the different stakeholders have about the tests and standards?

RAND was asked by the Air Force Director of Military Force Management Policy, Deputy Chief of Staff for Manpower, Personnel and Services to assist in evaluating issues emphasized in the IVT plan. Many of the issues are being systematically addressed by the AF-ESU (e.g., cost, number of test administrators required). Therefore, we concentrated our efforts on one particular issue in the IVT—concerns stakeholders have about the Operator–Prototype Test Battery (O-PTB). The issues identified as part of the IVT plan represent current priorities; however, other short- and long-term issues and concerns should be identified and may need to be subsequently addressed as implementation proceeds. To assist in identifying potential issues that may emerge as future concerns, we developed an evaluation framework, which presents a range of topics and research questions that can be revisited periodically to prioritize future research efforts.

Organization of This Report

The remainder of this report presents our approach and results for evaluating the implementation of the O-PTB and concludes with a proposed framework to guide future evaluation efforts. We followed these steps to address the primary research questions:

1. Identify relevant stakeholders.
2. Design evaluation instruments to address stakeholder reactions.
3. Collect and analyze the data.
4. Develop an evaluation framework for future evaluation efforts.

Chapter 2 addresses steps one and two by providing an overview of the key stakeholder groups we identified, specific topics relevant to each stakeholder group, and the evaluation tools we used to conduct a preliminary evaluation of the implementation of the O-PTB. Chapters 3, 4, and 5 address step three with an analysis and presentation of results from each stakeholder group: TACP and ALO operator perspective (Chapter 3), physical training leader (PTL) perspective (Chapter 4), and career field manager (CFM) perspective (Chapter 5). Chapter 6 describes a proposed evaluation framework with potential topics and research questions to guide future evaluation efforts. The final chapter provides a summary of the findings and offers recommendations for addressing potential concerns that may affect the successful implementation of continuation tests and standards for TACPs and ALOs.

2. Methodology: Initial Steps to Evaluate O-PTB Implementation

As indicated in Chapter 1, the O-PTB was generally found to be a strong predictor of task simulation performance. This finding offers initial validity evidence, which is critical to supporting implementation; however, successful implementation will require more than just scientific support. Specifically, stakeholders' acceptance will be an important prerequisite to future O-PTB adoption. Prior to this study, however, stakeholders' acceptance of the O-PTB had not been systematically captured. Stakeholder acceptance is one aspect of a larger "evaluation" framework, which we describe in Chapter 6. This framework captures multiple levels (e.g., individual, unit, career field) and can be extended to different contexts (e.g., recruiting, training).

The AF-ESU identified stakeholder acceptance among the list of questions to be addressed in its IVT plan ("What concerns do the different stakeholders have about the tests and standards?"). To help answer this question, we designed a mixed-method study to examine stakeholders' acceptance of the O-PTB. In this chapter, we describe our methodology. Specifically, we first identify which stakeholders are relevant to this stage of implementation (step one) and then discuss our process to design the evaluation instruments (step two). We conclude this chapter by describing the data-collection procedures and participant samples.

Establish Relevant Stakeholders

The Air Force intends to sequence the physical tests and standards implementation in three waves for TACPs and ALOs: (1) operators, (2) technical training students, and (3) recruits. This sequence was driven by the need to fill the immediate gap in the lack of an operator test for the TACP and ALO career fields. Consequently, the Air Force is in its first wave of implementation. Given that focus is on the operational community, we identified three relevant stakeholders:

- TACP and ALO operators, who will be required to take the tests and meet the specified standards
- CFMs, who will be responsible for addressing gaps in readiness levels for their specialty and for overseeing whether resource needs are being met across the career field
- PTLs, who will be responsible for administering and scoring the tests and providing training to other test administrators.

Understanding stakeholders' reactions to the new physical tests and standards is an important step to establishing the support needed for successful implementation. To address potential acceptance among each stakeholder group, our evaluation efforts are guided by the following questions:

- What are TACP and ALO operators' reactions to the new O-PTB?

- What are PTL's reactions to PF training administration and the new O-PTB?
- What are CFM's reactions to the new O-PTB and implementation of the O-PTB?

Design Evaluation Instruments to Address Stakeholder Reactions

To design our evaluation instruments, we followed a systematic process, as shown in Figure 2.1. We discuss this process in more detail in the following section.

Figure 2.1. Process to Design Evaluation Instruments

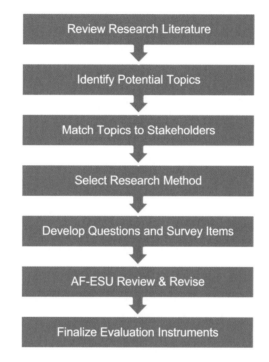

Applicant and Trainee Reactions Literatures Guided Our List of Potential Topics

We began our efforts by examining the applicant and trainee reactions literatures to identify possible topics. These reactions, respectively, refer to subjective experiences (e.g., positive attitudes) about the hiring process or training (e.g., Kirkpatrick, 1996; Ryan and Ployhart, 2000). Interestingly, these two literatures emerged from relatively distinct lines of research. Early work on the hiring process focused primarily on the organization's perspective of the recruiting and selection procedures; however, applicant (or job candidate) reactions began to receive attention because of a number of economic, legal, and psychological factors (Hülsheger and Anderson, 2009). For instance, meta-analytic evidence finds that applicants who perceive selection tools and processes to be fair and job-related are more likely to view the organization favorably, recommend the employer to others, and report stronger intentions to accept job offers than those who do not hold positive views (Hausknecht, Day, and Thomas, 2004). In contrast to the recent

7

interest in applicant reactions, trainee reactions have traditionally maintained a prominent role in training evaluation (Kirkpatrick, 1976) and are the most commonly assessed outcome of training (Patel, 2010). From our literature review (see Appendix B) and additional concerns relevant to PF testing (e.g., likelihood for injury), we identified a broad range of relevant topics. The most important of these topics are presented in Tables 2.1 through 2.3.

Topics Matched to Each Stakeholder Group

As is repeatedly emphasized and demonstrated in the literature (e.g., Derous and Schreurs, 2009; Salas et al., 2012), the purpose and context of evaluation must guide the design of evaluation instruments. Considering the specific stakeholder roles, we determined which topics were appropriate to a given stakeholder group (i.e., TACP and ALO operators taking the tests, PTLs administrating the tests, or CFMs implementing test policy). Certain stakeholders were uniquely positioned to provide certain information. For instance, TACP and ALO operators provided details about each PF test, PTLs addressed the test administration training, and CFMs conveyed information about potential challenges to implementation.

Selecting an Appropriate Research Method

Next, we selected the most suitable method for collecting feedback from each stakeholder group. Determining the most appropriate method was guided by the purpose and feasibility. For ALO and TACP operators and PTLs, we opted to use a survey methodology because we wanted to standardize response options to allow for direct comparisons on specific items and tests. For the majority of these questions, we used a five-point response scale (1 = *strongly disagree* and 5 = *strongly agree*). We also designed three open-ended questions to ensure sufficient opportunities for respondents to provide explanations or clarification for their responses. For CFMs, we decided to use a semistructured interview methodology because we wanted to capture the richness and nuances of their perspectives and to facilitate our understanding of their specific concerns.

Developing Question Items, Seeking AF-ESU Input, and Finalizing the Instruments

Taking into account the stakeholder group and research method, we developed specific questions and survey items. We then presented these questions and items to the AF-ESU for review. After incorporating their suggested recommendations, we finalized the instrument for administration. Table B.2 in Appendix B presents a comprehensive list of these items mapped onto each stakeholder group and each topic.

Primary Topic Areas Addressed

In Tables 2.1 through 2.3, we summarize the preceding steps by highlighting the most important topics for each stakeholder group and present a description of each topic (when needed) and an example item. Table 2.1 presents the primary topics for TACP and ALO operators, which concentrated on detailed information about each of the PF tests as well as more global perceptions of the O-PTB.

Table 2.1. Primary Topics for TACP and ALO Operators

Topic	Description	Example
Consistency of test administration	"Uniformity of content across test settings, in scoring, and in the interpretation of scores. Assurance that decision-making procedures are consistent across people and over time" (Bauer et al., 2012, p. 18–19)	[Specific PF test] was administered to all operators in the same way.
Knowledge of test performance relative to test standards	Extent to which performance on a test is readily apparent	I know how I performed on the [specific PF test] relative to the required standard.
Injury concerns related to the tests	The degree to which individuals are concerned about potential injury	I am concerned that I was injured or could have been injured while completing the [specific PF test].
Experienced levels of frustration in taking the tests	The extent to which individuals subjectively evaluate their test taking experience as negative (e.g., Alliger et al., 1997)	Taking the [specific PF test] was frustrating.
Global evaluations of the O-PTB		
Perceived utility	The extent to which individuals perceive testing results as useful (e.g., Alliger et al., 1997)	Knowing how well I did on the PF tests will help me improve my job-related physical capability.
Validity	"Extent to which a test either appears to measure the content of the job or appears to be a valid predictor of job performance" (Bauer et al., 2012, p. 18)	Doing well on this test battery means a person can perform the physical job duties of a TACP well.
Fairness	"Having adequate opportunity to demonstrate one's knowledge, skills and abilities in the testing situation" (Bauer et al., 2012, p. 18)	The testing process I just completed would be fair for all TACPs regardless of rank, age, stature, gender, or race/ethnicity.

Table 2.2 presents the two primary topics for PTLs, which focused on test administration training as well as more global perceptions of the O-PTB. We posed this latter topic to PTLs, as well as TACP and ALO operators, to compare their perspectives.[1]

---

[1] There are a total of 11 primary topics posed to stakeholders, but both PTLs and TACP and ALO operators responded to global evaluations; therefore, there were only ten unique topics.

Table 2.2. Primary Topics for PTLs

Topic	Description	Example
Quality of training provided to administer and score tests	Training component included in the test administrator training delivered to the PTLs	I know how to demonstrate and coach others on how to properly perform each test.
Global evaluations of the O-PTB		
Perceived utility	The extent to which individuals perceive testing results as useful (e.g., Alliger et al., 1997)	TACPs will find their test results useful for improving their own job-related PF.
Validity	"Extent to which a test either appears to measure the content of the job or appears to be a valid predictor of job performance" (Bauer et al., 2012, p. 18)	Doing well on this test battery means a person can perform the physical job duties of a TACP well.
Fairness	"Having adequate opportunity to demonstrate one's knowledge, skills and abilities in the testing situation" (Bauer et al., 2012, p. 18)	Operators could really show their physical abilities through this test battery.

Table 2.3 presents the four primary topics asked of the CFMs, which emphasized implementation-related issues. For this table, we did not include a topic description because the topic label and corresponding examples are sufficiently descriptive.

Table 2.3. Primary Topics for CFMs

Topic	Example
Current and future plans for test implementation	What plans does your career field have for further evaluation or implementation of an updated physical test battery?
Perceived benefits/challenges of implementing a new test battery	Do you see any benefits of updating the physical test battery (e.g., use different physical tests than the Physical Ability and Stamina Test (PAST)? Do you see any challenges?
Concerns with the specific recommended test battery	What concerns, if any, do you have about the O-PTB?
Specific barriers and concerns for test implementation such as time to administer, potential for injury, cost, fairness, and utility	What barriers or obstacles are likely to interfere with moving toward implementation of an updated test battery (not necessarily the one currently proposed prototype)?

Additional Topic Areas Addressed

As indicated earlier, we identified a wide range of topics for inclusion in our preliminary evaluation. Examples of other topics include the amount of information known prior to testing, a direct comparison to the PAST,[2] and the characteristics of the test administrator. The list of

---

[2] We use the PAST as a point of comparison because many of its test components (e.g., pull-ups, push-ups, sit-ups, 1.5-mile run) are commonly used across the BA specialties at different career stages, including for recruits, trainees, and operators. At a given career stage and/or for each BA specialty, however, there are variations in the standards (e.g., time, distance), ordering of test administration, and/or specific tests (e.g., inclusion of swimming tests).

topics, associated descriptions and relevant survey items are presented in Table B.2.[3] In Appendixes C and D, we present the actual survey completed by the TACP and ALO operators and PTLs, respectively.

Data-Collection Procedures

Surveys Administered to the TACP and ALO Operators and PTLs

The AF-ESU administered hard copies of the evaluation instruments during their installation visits. The AF-ESU collected surveys at 12 installations: Ft. Bliss, Ft. Campbell, Ft. Carson, Ft. Drum, Ft. Hood, Ft. Polk, Ft. Riley, Ft. Stewart, Ft. Wainwright, Joint Base Lewis-McChord, Pope Field, and Wheeler Army Field. The timing of survey administration was carefully considered. For the TACP and ALO operators, one section of the survey was administered immediately after the completion of each test. This section consisted of four questions (consistency, knowledge of performance, concern for injury, and frustration) specific to each test. The remainder of the survey, which consisted of general test perceptions and background characteristics such as rank, skill level, and years in the career field, was administered after all PF tests were completed. For PTLs, questions evaluating training were administered immediately after the test-administration training. The remaining portions of the survey were administered upon the conclusion of the PTL test administration to ALOs and TACPs. After data collection, the AF-ESU securely transferred completed surveys to RAND for data entry and analyses.

Semistructured Interviews Conducted with the BA CFMs

We conducted three semistructured interviews with CFMs from the following career fields: (1) ALO and TACP; (2) PJ, CCT, and CRO; and (3) SOWT. The semistructured interviews lasted approximately one hour. We recorded notes from each of the meetings and later analyzed this information for emerging themes.

Survey Sample Characteristics

Table 2.4 describes the respondent characteristics for the 198 ALO and TACP operators that completed the survey. These individuals were mostly enlisted (E3 to E5) and represented units from 12 different installations.

---

[3] To help the Air Force plan for future evaluation efforts, we also developed similar instruments for other stakeholder groups, including prospective TACP candidates, field developers, and TACP students. Although these have not yet been used in the field, they were developed to help the Air Force plan for its future evaluation efforts.

Table 2.4. TACP and ALO Respondent Characteristics

Variable	Levels	N	%	Cumulative %
Rank	E2	1	0.5	0.5
	E3	30	15.2	15.7
	E4	49	24.8	40.4
	E5	38	19.2	59.6
	E6	14	7.1	66.7
	E7	11	5.6	72.2
	E8	4	2.0	74.2
	E9	2	1.0	75.3
	O1	3	1.5	76.8
	O2	6	3.0	79.8
	O3	6	3.0	82.8
	O4	2	1.0	83.8
	O5	1	0.5	84.4
	Missing	31	15.7	100.0
Deployments[a]	0	84	42.4	42.4
	1 to 3	53	26.8	69.2
	4 to 5	16	8.1	77.3
	> 5	16	8.1	85.3
	Missing	29	14.7	100.0
Operations[a]	1 to 25	82	41.4	41.4
	26 to 50	44	22.2	63.6
	51 or more	39	19.7	83.3
	Missing	33	16.7	100.0
Height	61 to 67 inches	30	15.2	15.2
	68 to 71 inches	90	45.5	60.6
	72 to 77 inches	50	25.2	85.9
	Missing	28	14.1	100.0
Weight	< 150 lbs	6	3.0	3.0
	151 to 180 lbs	64	32.3	35.4
	181 to 200 lbs	58	29.3	64.6
	201 to 230 lbs	37	18.7	83.3
	> 230 lbs	3	1.5	84.8
	Missing	30	15.2	100.0
Location	Ft. Bliss	11	5.6	5.6
	Ft. Campbell	35	17.7	23.2
	Ft. Carson	3	1.5	24.8
	Ft. Drum	35	17.7	42.4
	Ft. Hood	10	5.0	47.5
	Ft. Polk	6	3.0	50.5
	Ft. Riley	10	5.0	55.6

Variable	Levels	N	%	Cumulative %
	Ft. Stewart	5	2.5	58.1
	Ft. Wainwright	20	10.1	68.2
	JBLM	18	9.1	77.3
	Pope	17	8.6	85.9
	Wheeler	28	14.1	100.0
	All	198	100.0	100.0

NOTE: JLBM = Joint Base Lewis-McChord.
[a] Subsequent input from the operators indicated confusion regarding the specific definition of deployments and operations; therefore, we would suggest further refinement in defining categories for future iterations. Operations were intended to capture the number of specific mission-related training exercises and actual engagements, whereas deployments were meant to capture the number of times when individuals engage in extended periods of duty away from their home station.

Table 2.5 describes the respondent characteristics for the 135 PTLs that completed the survey. The majority of these individuals were enlisted (E4 or E5) and had few deployments (three or fewer).

Table 2.5. PTL Test Administrator Characteristics

Variable	Levels	N	%	Cumulative %
Rank	E2	2	1.5	1.5
	E3	9	6.7	8.2
	E4	36	26.7	34.8
	E5	41	30.4	65.2
	E6	13	9.6	74.8
	E7	6	4.4	79.3
	E8	1	0.7	80.0
	E9	1	0.7	80.7
	O1	1	0.7	81.5
	O2	2	1.5	83.0
	O3	4	3.0	85.9
	O4	1	0.7	86.7
	Missing	18	13.3	100.0
Deployments	0	50	37.0	37.0
	1 to 3	49	36.3	73.3
	4 to 5	9	6.7	80.0
	> 5	10	7.4	87.4
	Missing	17	12.6	100.0
Operations	1 to 25	49	36.3	36.3
	26 to 50	39	28.9	65.2
	51 or more	7	5.2	70.4
	Missing	40	29.6	100.0
	All	135	100.0	
Height	61 to 67 inches	22	16.3	16.3
	68 to 71 inches	53	39.3	55.6

Variable	Levels	N	%	Cumulative %
	72 to 77 inches	44	32.6	88.2
	Missing	16	11.8	100.0
Weight	< 150 lbs	3	2.2	2.2
	151 to 180 lbs	41	30.4	32.6
	181 to 200 lbs	45	33.3	65.9
	201 to 230 lbs	28	20.7	86.7
	> 230 lbs	2	1.5	88.1
	Missing	16	11.8	100.0
Location	Ft. Bliss	8	5.9	5.9
	Ft. Campbell	9	6.7	12.6
	Ft. Carson	15	11.1	23.7
	Ft. Drum	19	14.1	37.8
	Ft. Hood	18	13.3	51.1
	Ft. Polk	9	6.7	57.8
	Ft. Riley	6	4.4	62.2
	Ft. Stewart	5	3.7	65.9
	Ft. Wainwright	8	5.9	71.8
	JBLM	10	7.4	79.3
	Pope	12	8.9	88.2
	Wheeler	16	11.8	100.0
	All	135	100.0	100.0

Conclusion

This chapter presented the methodology for our preliminary evaluation of the new O-PTB. Based on the evaluation framework (Appendix E), we identified research questions about the perceptions of the O-PTB and/or implementation of the O-PTB for ALO and TACP operators, PTLs, and CFMs. We then relied on a systematic process to address these research questions, which involved identifying potential topics by reviewing the literature, matching appropriate topics and research designs to the three stakeholder groups, further tailoring and/or creating items to assess topics of interest, and collecting corresponding data. The next three chapters will discuss the results of these evaluation efforts.

3. Test-Taker Perspectives from TACPs and ALOs

This chapter presents detailed results on the perceptions of tests and test implementation from TACPs and ALOs. In the first section, we present TACP and ALO responses to a standard set of questions for each test, with the exception of the 1.5-mile run. The 1.5-mile run test was not administered as part of the AF-ESU's implementation trips because the Air Force already conducts this test regularly as part of the Tier I Air Force–wide fitness test.

Survey Responses

The first section of survey results presents the distribution of responses for each of the nine tests on the following four statements:

1. The [O-PTB subtest] was administered to all operators in the same way.
2. I know how I performed on the [O-PTB subtest] relative to the required standard.
3. I am concerned that I was injured or could have been injured while completing the [O-PTB subtest].
4. Taking the [O-PTB subtest] was frustrating.

In most cases, we observed little variability in responses across these four survey items for each test. If variation was present, as indicated by a value of 20 percent or greater at both ends of the scale (e.g., 20 percent disagree/strongly disagree, 60 percent neither, 20 percent agree/strongly agree), we conducted follow-up analyses by comparing responses across installation and demographic variables (e.g., rank, height) to explore potential explanations for the observed variation. To further support the identification of possible explanations for observed variation, we reviewed open-ended responses. Although many of the open-ended responses indicated general positive support for the overall test battery, other comments clustered around specific concerns and recommendations. These most frequently cited concerns are presented in the following sections. A full list of concerns organized by theme for each open-ended question is provided in Appendix F.

Consistency of Test Administration

The tests were administered in the same order for each participant, and standardized rest periods were provided between each test.[1] Overall, the pattern of TACP and ALO responses indicated consistent agreement that each test was administered consistently. Specifically, over

---

[1] The ESU established the order of the test battery following standard exercise science testing principles (i.e., tests administered in order of recovery time required following each test). That is, the most demanding test (1,000-meter row ergometer) was administered last.

90 percent of respondents agreed or strongly agreed that each was consistently administered (see Figure 3.1). Furthermore, only two tests, Pull-Ups and Extended Cross Knee Crunch, had less than 95 percent agreeing or strongly agreeing that these tests were administered consistently (94 and 92 percent, respectively).

Figure 3.1. The O-PTB Was Administered to All Operators in the Same Way

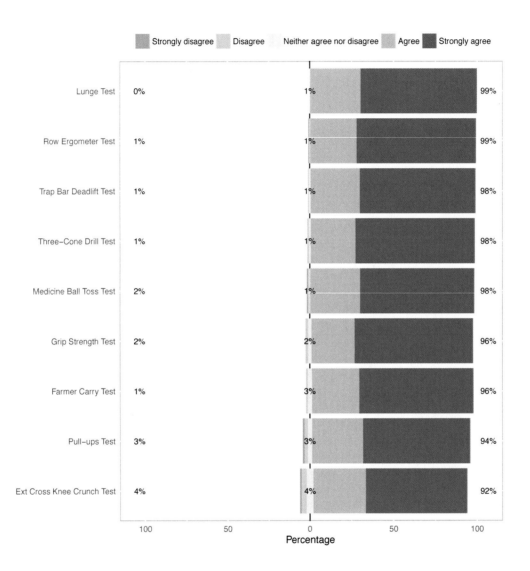

Knowledge of Performance Relative to Standard

All operators were provided with information about the required standards for each test. Therefore, this item emphasizes whether operators could determine how well they performed or were provided with sufficient feedback on their performances. Overall, TACP and ALOs indicated that they knew how well they performed on each test relative to the required standard. Only the Extended Cross Knee Crunch yielded less than 90 percent of responses agreeing or strongly agreeing, with 9 percent of respondents indicating they neither agreed nor disagreed with the statement (see Figure 3.2).

Figure 3.2. I Know How Well I Performed Relative to the Required Standard

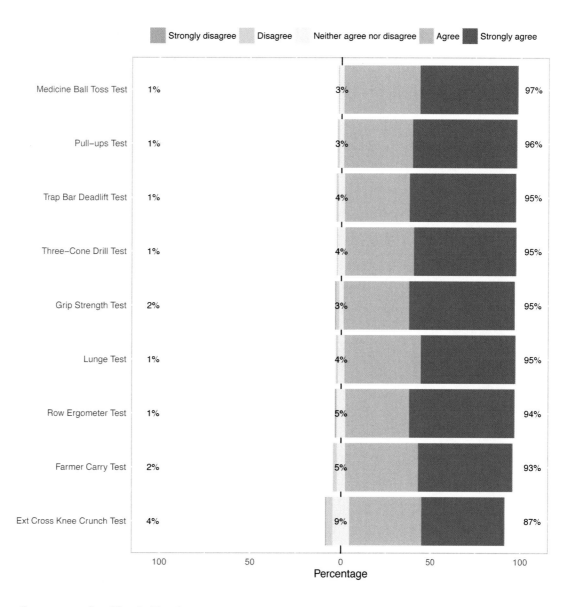

Injury Concerns for Each Test

Overall, TACP and ALO respondents indicated few injury concerns across the majority of the tests. However, 20 percent of respondents indicated an injury concern for the Trap Bar Deadlift (see Figure 3.3), which met our established threshold for further exploratory analysis to identify potential explanations for these concerns.

17

Figure 3.3. I Am Concerned That I Was Injured or Could Have Been Injured While Completing the Subtest

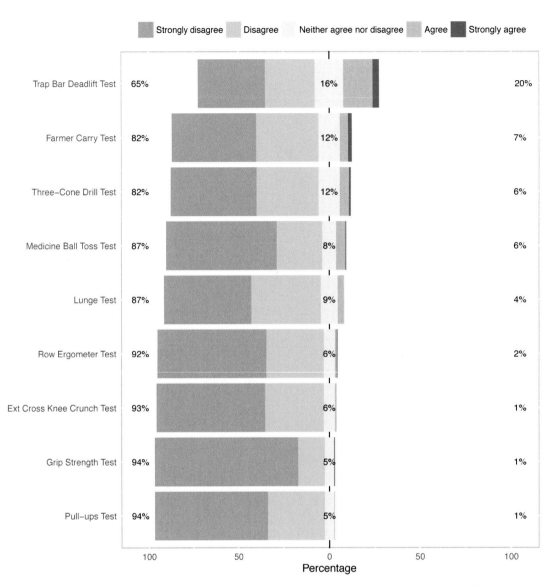

First, we conducted an analysis of variance test to determine if there were significant differences in injury concerns across tests. Specifically, we found there was a significant effect of test on response agreement to an injury concern at the $p < .05$ level across the nine tests $[F (8, 197) = 35.64, p < .05]$. Further analysis (i.e., post hoc comparisons) indicated significantly more respondents with injury concerns for the Trap Bar Deadlift compared with other tests in the test battery (see Table 3.1).

However, it is important to note that the number of respondents indicating a concern about injury during the Trap Bar Deadlift is still relatively low compared with the number indicating no concern. Nonetheless, we conducted additional analyses to identify whether concerns were clustered around a few installations (see Figure 3.4) or whether concerns were associated with

18

background characteristics, such as height, weight, and deployment experience. These analyses revealed no significant patterns that would differentiate TACPs and ALOs who have concerns compared with those who do not have concerns about injuries resulting from the Trap Bar Deadlift.

Table 3.1. Injury Concern Test Comparisons

Test Comparison	Difference	Lower	Upper	p-Value Adjusted
Trap Bar to Grip Strength	**0.95**	0.74	1.15	0.00
Trap Bar to Pull-Up	**0.78**	0.58	0.99	0.00
Trap Bar to Cross Knee	**0.74**	0.53	0.94	0.00
Trap Bar to Row Ergometer	**0.72**	0.52	0.93	0.00
Trap Bar to Medicine Ball	**0.64**	0.43	0.84	0.00
Trap Bar to Lunge	**0.54**	0.33	0.74	0.00
Three Cone to Grip Strength	**0.50**	0.30	0.70	0.00
Trap Bar to Three Cone	**0.45**	0.24	0.65	0.00
Trap Bar to Farmer Carry	**0.42**	0.21	0.62	0.00
Lunge to Grip Strength	**0.41**	0.20	0.61	0.00
Three Cone to Pull-Up	**0.34**	0.14	0.54	0.00
Farmer Carry to Cross Knee	**0.32**	0.12	0.52	0.00
Medicine Ball to Grip Strength	**0.31**	0.11	0.51	0.00
Three Cone to Cross Knee	**0.29**	0.09	0.49	0.00
Three Cone to Row Ergometer	**0.28**	0.08	0.48	0.00
Row Ergometer to Grip Strength	**0.22**	0.02	0.42	0.02
Lunge to Cross Knee	0.20	0.00	0.40	0.06
Three Cone to Medicine Ball	0.19	−0.01	0.39	0.09
Pull-Up to Grip Strength	0.16	−0.04	0.37	0.25
Medicine Ball to Cross Knee	0.10	−0.10	0.31	0.83
Three Cone to Lunge	0.09	−0.11	0.29	0.90
Row Ergometer to Pull-Up	0.06	−0.14	0.26	0.99
Row Ergometer to Cross Knee	0.01	−0.19	0.22	1.00
Three Cone to Farmer Carry	−0.03	−0.23	0.17	1.00
Pull-Up to Cross Knee	−0.05	−0.25	0.16	1.00
Row Ergometer to Medicine Ball	−0.09	−0.29	0.11	0.91
Medicine Ball to Lunge	−0.10	−0.30	0.10	0.85
Lunge to Farmer Carry	−0.12	−0.32	0.08	0.66
Pull-Up to Medicine Ball	−0.15	−0.35	0.06	0.37
Row Ergometer to Lunge	−0.19	−0.39	0.02	0.10
Grip Strength to Cross Knee	**−0.21**	−0.41	0.00	0.04
Medicine Ball to Farmer Carry	**−0.22**	−0.42	−0.02	0.02
Pull-Up to Lunge	**−0.25**	−0.45	−0.04	0.01
Row Ergometer to Farmer Carry	**−0.31**	−0.51	−0.11	0.00
Pull-Up to Farmer Carry	**−0.37**	−0.57	−0.16	0.00
Grip Strength to Farmer Carry	**−0.53**	−0.73	−0.33	0.00

NOTE: Shaded rows and differences in **bold** are significant at p-value less than .05. A positive difference value indicates that the first test raised greater injury concern compared with the second test. The difference is the mean difference between the two tests on the 1–5 point agreement scale.

19

Figure 3.4. Responses for Trap Bar Deadlift Injury Concern by Installation

20

Although no patterns emerged across subgroups in relation to injury concerns for the Trap Bar Deadlift, respondents' open-ended comments primarily emphasized the ability to use proper form and technique when performing the lift. Some of the respondents provided specific suggestions to address potential injury concerns, including providing sufficient training on proper form, providing opportunities to practice, and using a weight belt when lifting.

Frustration Experienced

Tests that produced the most frustration included both the Extended Cross Knee Crunch (41 percent) and the Medicine Ball Toss (21 percent) (see Figure 3.5). We conducted additional analyses to determine if any significant patterns in frequency of frustration emerged by installation and background characteristics for these two tests. First, we conducted an analysis of variance test to determine if there were significant differences in frustration concerns across tests. Specifically, we found there was a significant effect of test on response agreement to frustration at the $p < .05$ level across the nine tests [$F (8, 196) = 29.68, p < .05$]. Further analysis (i.e., post hoc comparisons) indicated significantly more frustration for the Extended Cross Knee Crunch and the Medicine Ball Toss compared with other tests in the test battery (see Table 3.2).

Figure 3.5. Taking the O-PTB Was Frustrating

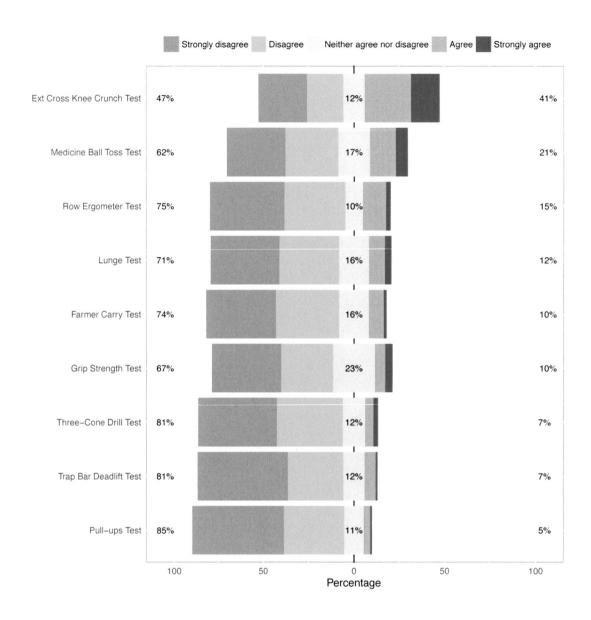

Table 3.2. Frustration Concern Test Comparisons

Test Comparison	Difference	Lower	Upper	p-Value Adjusted
Pull-Up to Cross Knee	**−1.14**	−1.42	−0.86	0.00
Trap Bar to Cross Knee	**−1.07**	−1.35	−0.79	0.00
Three Cone to Cross Knee	**−0.98**	−1.26	−0.71	0.00
Farmer Carry to Cross Knee	**−0.85**	−1.13	−0.57	0.00
Row Ergometer to Cross Knee	**−0.82**	−1.10	−0.54	0.00
Lunge to Cross Knee	**−0.77**	−1.05	−0.49	0.00
Grip Strength to Cross Knee	**−0.76**	−1.03	−0.48	0.00
Pull-Up to Medicine Ball	**−0.64**	−0.92	−0.36	0.00
Trap Bar to Medicine Ball	**−0.57**	−0.85	−0.29	0.00
Medicine Ball to Cross Knee	**−0.50**	−0.78	−0.22	0.00
Three Cone to Medicine Ball	**−0.48**	−0.76	−0.21	0.00
Pull-Up to Grip Strength	**−0.38**	−0.66	−0.11	0.00
Pull-Up to Lunge	**−0.37**	−0.65	−0.09	0.00
Trap Bar to Grip Strength	**−0.32**	−0.59	−0.04	0.01
Row Ergometer to Medicine Ball	**−0.32**	−0.60	−0.04	0.01
Trap Bar to Lunge	**−0.30**	−0.58	−0.02	0.02
Pull-Up to Farmer Carry	**−0.29**	−0.56	−0.01	0.04
Trap Bar to Row Ergometer	−0.25	−0.53	0.03	0.12
Three Cone to Grip Strength	−0.23	−0.51	0.05	0.20
Trap Bar to Farmer Carry	−0.22	−0.50	0.06	0.26
Three Cone to Lunge	−0.21	−0.49	0.06	0.28
Three Cone to Row Ergometer	−0.16	−0.44	0.11	0.66
Three Cone to Farmer Carry	−0.13	−0.41	0.14	0.86
Trap Bar to Three Cone	−0.09	−0.36	0.19	0.99
Row Ergometer to Grip Strength	−0.07	−0.34	0.21	1.00
Row Ergometer to Lunge	−0.05	−0.33	0.23	1.00
Lunge to Grip Strength	−0.01	−0.29	0.26	1.00
Row Ergometer to Farmer Carry	0.03	−0.25	0.31	1.00
Trap Bar to Pull-Up	0.07	−0.21	0.34	1.00
Lunge to Farmer Carry	0.08	−0.20	0.36	0.99
Grip Strength to Farmer Carry	0.10	−0.18	0.37	0.98
Three Cone to Pull-Up	0.15	−0.12	0.43	0.74
Medicine Ball to Grip Strength	0.26	−0.02	0.53	0.10
Medicine Ball to Lunge	**0.27**	−0.01	0.55	0.06
Row Ergometer to Pull-Up	**0.32**	0.04	0.59	0.01
Medicine Ball to Farmer Carry	**0.35**	0.07	0.63	0.00

NOTE: Shaded rows and differences in **bold** are significant at p-value less than .05. A positive difference value indicates that the first test raised greater injury concern compared to the second test. The difference is the mean difference between the two tests on the 1–5 point agreement scale.

Additional Analysis for Extended Cross Knee Crunch

Additional analyses were conducted to identify whether frustration concerns were clustered around a few installations (see Figure 3.6) or whether concerns were associated with background characteristics. Although a visual interpretation across installations shows that respondents in some locations (e.g., Ft. Polk) were less likely to indicate frustration with the Extended Cross Knee Crunch, these differences are not significant. Similar analyses comparing responses by demographic variables (e.g., height, weight, deployment experience) also revealed no significant patterns that would differentiate TACPs and ALOs who have concerns compared with those who do not have frustration concerns with the Extended Cross Knee Crunch.

Figure 3.6. Responses for Extended Cross Knee Crunch by Installation

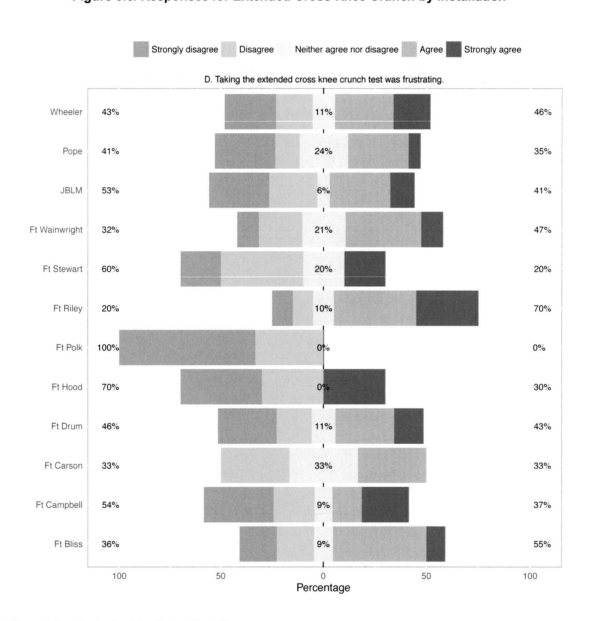

Additional Analysis for Medicine Ball Toss

Additional analyses were conducted to identify whether frustration concerns were clustered around a few installations (see Figure 3.7) or whether concerns were associated with background characteristics. Although a visual interpretation across installations shows that respondents in some locations (e.g., Ft. Polk and Ft. Carson) were less likely to indicate frustration with the Medicine Ball Toss, these differences were not significant. Similar analyses comparing responses by demographic variables, however, revealed that TACPs and ALOs who are heavier and taller were significantly less likely to indicate frustration compared with lighter and shorter TACPs and ALOs (see Figure 3.8). This finding is further supported by a few open-ended comments suggesting that the Medicine Ball Toss appears to favor taller individuals. Other analyses indicated significantly less frustration among more experienced operators as measured by years in the career field ($r = -.23$, $p < .05$) and number of deployments ($r = -.22$, $p < .05$). Open-ended comments indicated that some respondents felt that the Medicine Ball Toss required a lot of technique and practice to do well, with some stating that the test measures more skill/technique than strength/power. Others indicated frustration in not being able to move or raise their feet during the toss.

Figure 3.7. Medicine Ball Toss Frustration by Installation

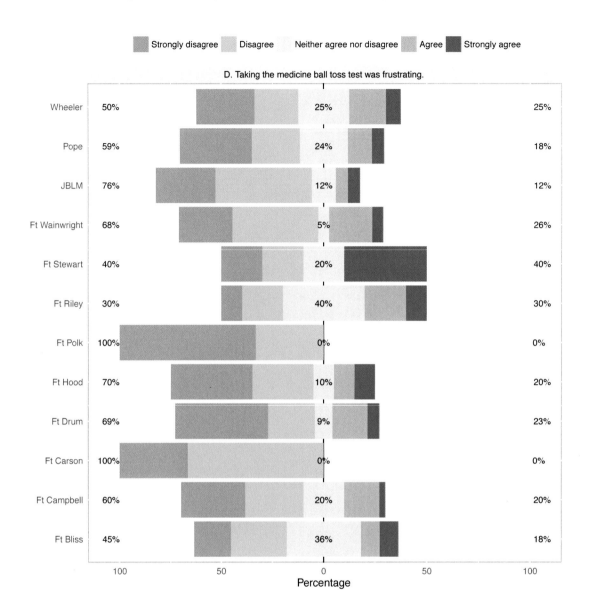

Figure 3.8. Medicine Ball Toss Frustration by Weight and Height

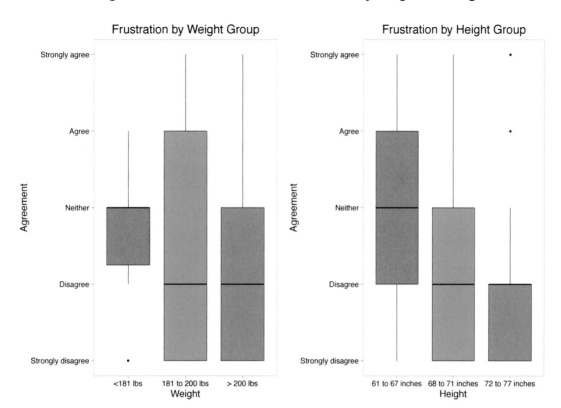

Global Evaluations of Prototype Tests and Test Administration

Several items were used to address operators' perspectives on the perceived utility of the test battery, in addition to perceptions of how well they think the test battery measures important abilities for their job (i.e., face validity). Overall, the general perceptions suggest that the test battery will be positively received by TACPs and ALOs, with 84 percent of respondents indicating that knowing how well they performed on this test battery will help improve job-related physical capabilities, and 74 percent indicating that this test measures the abilities required of a TACP (see Figure 3.9). Open-ended comments were also consistent with these ratings. For example, one respondent stated, "this test encompasses the strength, endurance, and stamina needed to perform the job at the required physical level."

Although most responses present favorable perceptions, 18 percent indicated that they disagreed or strongly disagreed that the test would be fair for all TACPs regardless of rank, age, stature, gender, or race/ethnicity. Respondents' open-ended comments suggested that the tests and standards might not be relevant to all TACPs given that some assignments do not require any (or as much) physical effort to perform their job tasks. For example, one respondent stated, "in reality the jobs are different at the higher ranks and . . . echelons."

27

Figure 3.9. TACP and ALO Global Perceptions

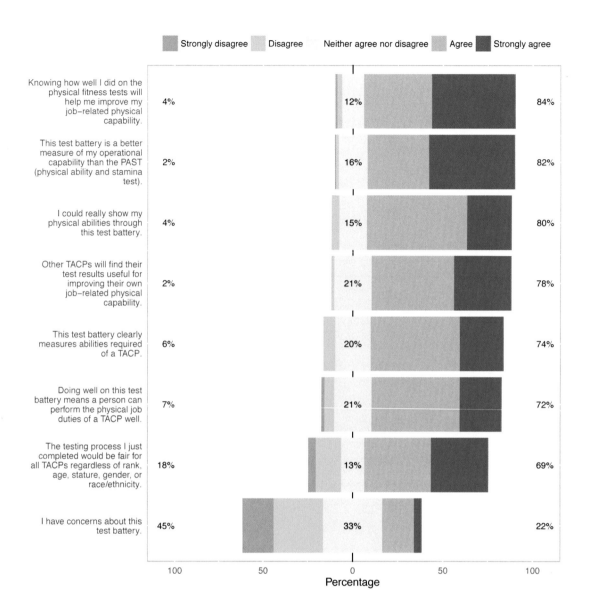

In addition to some concerns about fairness, 22 percent of respondents indicated that they had concerns about this test battery. Open-ended comments reiterated concerns previously discussed, including potential injuries during specific tests (e.g., Trap Bar Deadlift) and the importance of form and technique when taking the test (e.g., Medicine Ball Toss). However, other respondents raised concerns about the testing environment, emphasizing a range of potential issues that could affect the standardization of testing. Specifically, respondents mentioned the weather, surface conditions for running tests (e.g., Three Cone Drill), and potential space constraints. Space constraints were acknowledged by the chief of the AF-ESU, who suggested that changes might need to be made to the Farmer's Carry because it was particularly difficult to find 100 yards of flat surface at each location. One option being considered by the AF-ESU is identifying a shorter space (e.g., 25 yards) and having test participants carry back and forth until the full distance of 100 yards is met.

Several other questions directly asked respondents about how they were treated during the test administration (see Figure 3.10). Responses to these items were very positive, with 99 percent indicating that they were treated with respect, 97 percent indicating that questions were answered in a straightforward manner, and 90 percent indicating they were satisfied with the feedback received about their test results. One question was also designed to address operators' preference for who should be a test administrator. Seventy-one percent felt it was important the test administrators should be other TACPs.

Figure 3.10. TACP and ALO Perspectives on Test Administration

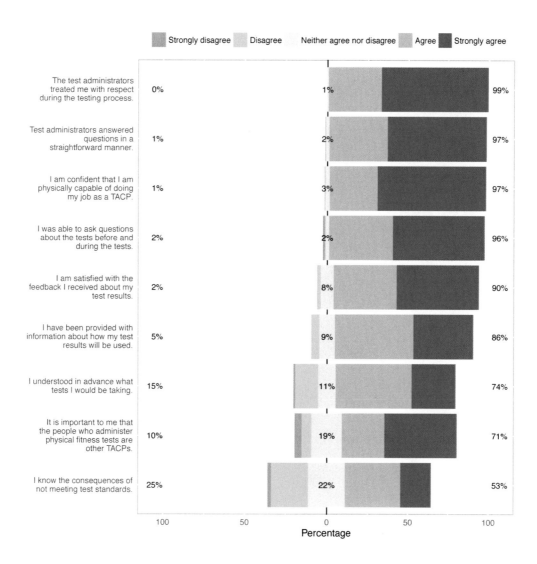

29

4. Summary of PTL Perspectives

In this chapter, we present results from the PTL perspective. The PTLs responded to survey items similar to those provided to TACPs and ALOs, in addition to items about the test administrator training they each received. PTLs were also provided with the opportunity to provide open-ended comments, which are summarized in Appendix G. The results are presented first for PTL responses to the training, followed by PTL global perceptions of the test battery, and finally for PLT perceptions about the utility of the test battery.

PTL Training

As described in Chapter 2, the AF-ESU provided training to PTLs on how to properly administer and score each test. As indicated in Figure 4.1, almost all (99 to 100 percent) of the PTLs agreed or strongly agreed with each survey item about training, with the exception of one item: "knowing the procedures to protect, store, and transfer the data." Follow-on discussions with the TACP CFM and the chief of the AF-ESU indicated that this item was confusing to some of the PTLs because they did not necessarily perform these functions. Overall, the PTL responses to the training clearly suggest that the training provided was effective.

Figure 4.1. PTL Responses to Test Administrator Training

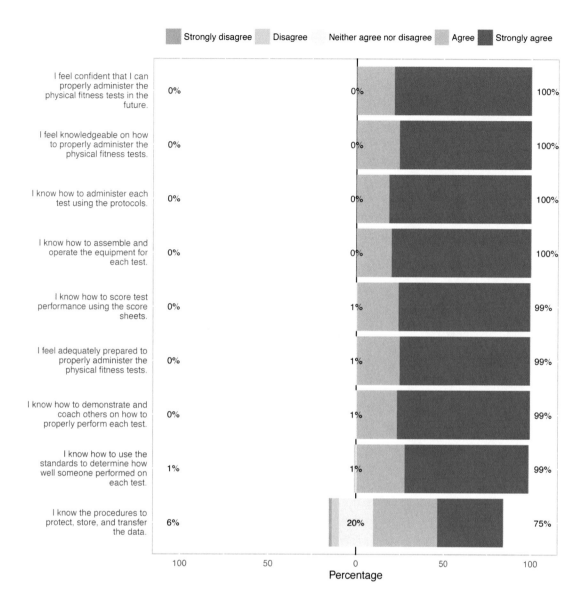

PTL Global Evaluations of the Test Battery

PTLs also provided responses to general questions about the prototype test battery (see Figure 4.2). Overall, these responses were also favorable. Ninety-three percent of respondents agreed or strongly agreed that each test was administered to all operators in the same way. Eighty percent of PTLs indicated that the test battery would be fair to all TACPs regardless of rank, age, stature, gender, or race/ethnicity. Some of those who indicated concerns about fairness provided open-ended comments suggesting that fitness requirements may not be relevant across the entire career lifecycle of a TACP—that is, TACPs may be assigned to a position, such as a staff position, in which fitness is less important. Other open-ended comments suggest that some of the frustration observed in taking specific tests (e.g., Extended Cross Knee Crunch) may be related to lack of familiarity with a new test and the desire to perform well. In some cases, PTLs as well as TACPs and ALOs indicated general frustration with the required standard for a test or with the distance between different scale scores (e.g., the number of additional pull-ups to move from a scale score of 2 to a scale score of 3 is three additional pull-ups).

Figure 4.2. PTL Global Perceptions of the Test Battery

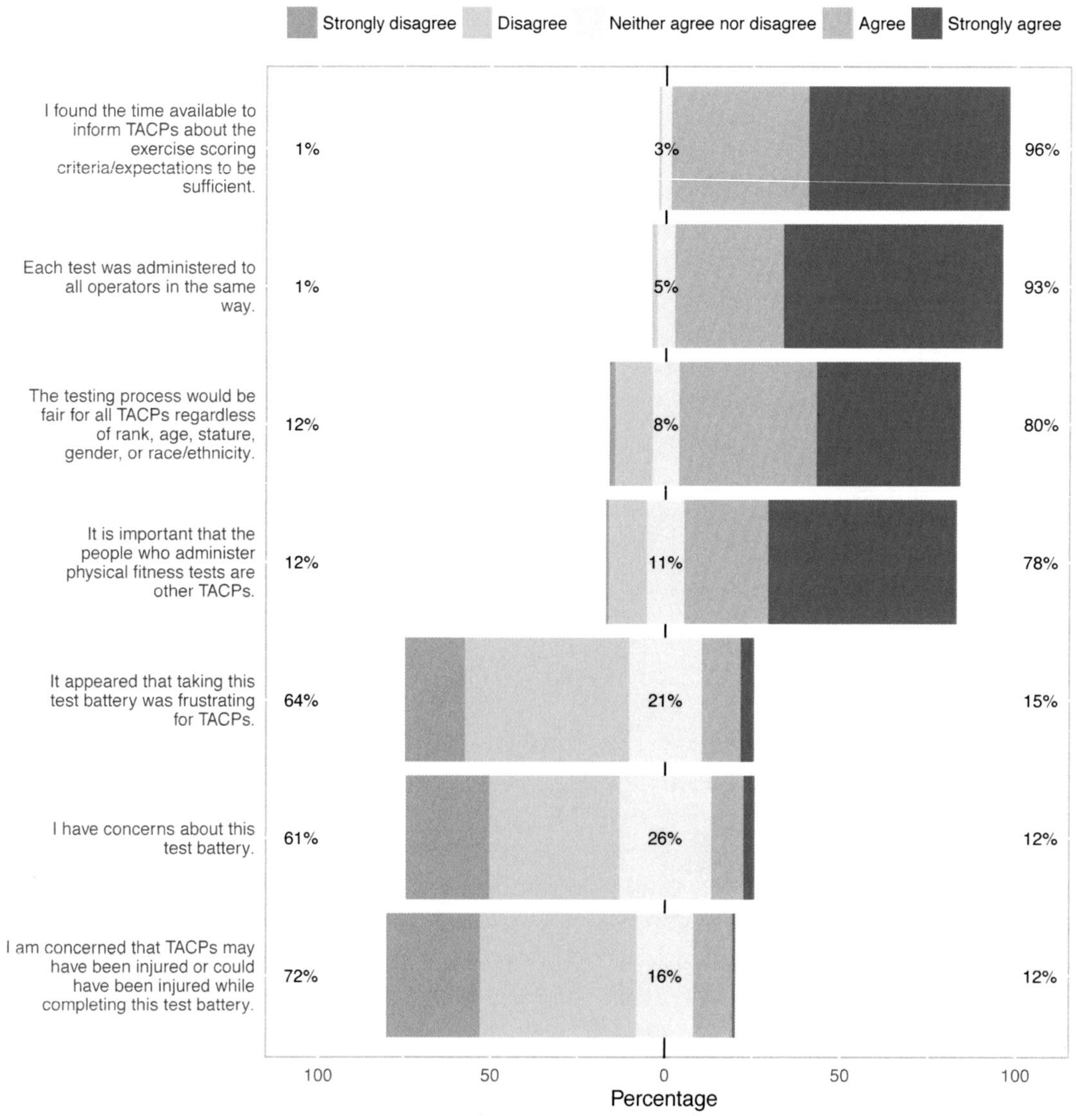

PTL Perceptions on the Utility of the Test Battery

The final set of survey items for PTLs focused on the overall utility of the test battery (see Figure 4.3). Responses indicated very positive support, with 97 percent agreeing or strongly agreeing that the test battery is a better measure of operational capabilities than the PAST. Furthermore, 94 percent agreed or strongly agreed that operators could really show their physical abilities through this test battery. Ninety-two percent indicated that TACPs would find their test results useful for improving their job-related physical capabilities.

Figure 4.3. PTL Perceptions on the Utility of the Test Battery

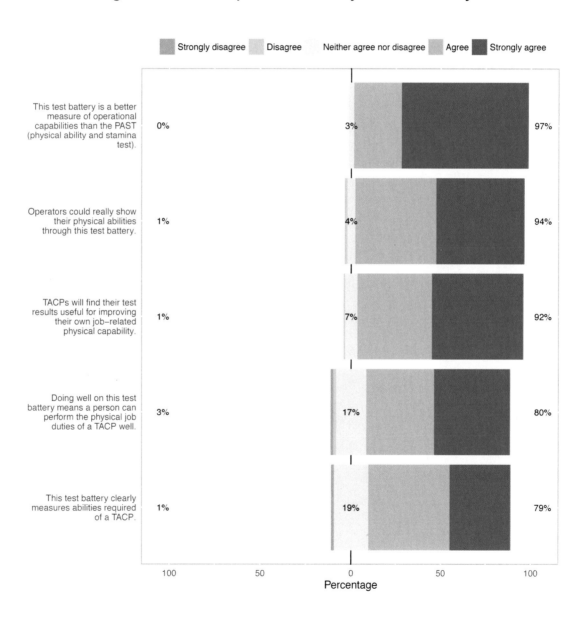

5. Summary of CFMs' Perspectives

To gain a better understanding of the BA CFMs' concerns about test implementation, we held three separate meetings with individual and small groups of CFMs, which included CFMs and career field representatives from TACP and ALO; CRO, PJ, and STO; and SOWT. One CFM (CCT) was unavailable for these meetings. Although only the TACP and ALO specialties have moved forward with implementing the new operator test battery, we decided to include other BA CFMs to provide a more comprehensive understanding of the issues, concerns, and barriers influencing the successful implementation of an updated physical test battery. Given the differences in current and planned efforts to implement an updated physical test battery, we present our discussion of the CFM feedback for each topic area. We split our discussion into two groupings: TACP/ALO CFMs and other BA CFMs. Several questions were presented to address the following topics:

1. current and future plans for test implementation
2. perceived benefits of implementing a new test battery
3. perceived challenges or drawbacks
4. concerns with the specific recommended test battery
5. specific barriers to and concerns about test implementation, such as time to administer, potential for injury, cost, fairness, and utility.

We begin our presentation of the CFM feedback by organizing comments provided by CFMs for each topic into a summary table (see Table 5.1). We expand on this table in the following sections with more-detailed descriptions of the comments provided by CFMs.

Table 5.1. Summary of CFMs' Perspectives on the Implementation of Tests and Standards

Topic	TACP/ALO CFMs	Other BA CFMs
Plans for test implementation	• Focus is on implementation for current operators • Future efforts will consider implementation for technical training students and future recruits	• General interest and will monitor how well tests and standards are implemented for TACP/ALO • No concrete plans to change tests and standards
Perceived benefits	New O-PTB is more comprehensive and addresses deficiencies in PAST by measuring job-related agility	New O-PTB addresses deficiency in PAST by measuring muscular power
Perceived challenges	• Tests and standards may not be equally relevant for TACPs/ALOs at all organizational levels (e.g., staff positions) • Fear of the unknown (e.g., potential career consequences if operators do not meet standards)	• Administration is perceived to take more time and will be more difficult to manage compared with the test battery currently used for their specialties • Additional equipment, which will require more money to purchase and maintain
Concerns with recommended test battery	Concern that commanders may overemphasize or underemphasize role of fitness for specialties; emphasized importance of balanced integration of fitness-related workouts and testing	• Insufficient communication on how final ten physical tests were selected from 39 considered in the validation study • Questioned value added by proposed tests and standards
Other comments	Communication often gets lost over time, and not all operators recall steps taken to develop recommended tests and standards	Recognize the value in making improvements, but do not agree that the O-PTB is correct solution

Plans for Implementation

TACP and ALO CFMs

TACP and ALO CFMs, in conjunction with AF-ESU, have been actively collecting data from operators to determine the rate of success on the proposed test standards and to identify any other potential issues (e.g., equipment) that need to be addressed prior to full implementation. The AF-ESU is taking the lead on evaluating several other questions (e.g., do successful test takers perform job tasks better or more efficiently than poor test takers?), as outlined in the implementation plan. The focus is currently on the operator test battery, but the career fields expect to adopt an updated test battery with different tests and standards for recruits and trainees in the future.

Other BA CFMs

There will be a minimum of two years before any changes to the current test batteries will be considered. There is a general interest in knowing how well the proposed test battery will work for TACPs and ALOs once implemented, but the results would need to be interpreted carefully because other BA specialties have different job requirements than TACPs and ALOs.

Benefits of a New Physical Test Battery

TACP and ALO CFMs

TACP and ALO CFMs indicated that the proposed test battery covers areas that are deficient in the current entry-level PAST, such as agility. In addition to a more comprehensive test battery, CFMs indicated that education about individual strengths and weaknesses will improve because strength and conditioning coaches will be using results to guide feedback to operators. The CFMs also acknowledged the benefits of using a composite score that does not cater to one specific body type, as operators need to be well rounded across all fitness components and cannot focus on one specific test. Specific minimums across all tests may still be necessary to ensure operators can perform the full range of CPTs.

Other BA CFMs

There is broad recognition that the PAST is deficient and primarily emphasizes cardiovascular endurance. An updated test battery could address some of these deficiencies through adding tests that measure muscular power. In general, CFMs indicated that an updated test should be more holistic, taking into consideration how best to test and train for the mission.

Challenges or Drawbacks of a New Physical Test Battery

TACP and ALO CFMs

TACP and ALO CFMs indicated that the proposed test battery standards might not be equally relevant to all positions and assignments. Specifically, the standards fit the brigade combat team and below well, but may not be as applicable to TACPs and ALOs who have received a certain level of leadership (e.g., at the division level or higher). Another concern is that officers, who are sometimes assigned to an ALO billet for a couple of years, may not have the fitness levels of an ALO nor be held to the ALO PF standards. This issue raises an important question for the career field: Which CPTs should an officer who is assigned as an ALO still be expected to perform if he or she is not held to the same standards as other ALOs?

Another important concern, which can be readily addressed through additional communication, is the concern among the operators of the potential consequences if they cannot meet the standards. (It appears that operators who have had a chance to take the tests are much less concerned about this.) Some concerns emphasize the importance of communicating the process and science behind the new test battery; some operators do not understand why there are minimum scores required for each test in addition to a composite score across all tests. Related to the scoring system, some of the career field leaders want to have the tests scored on a 0–100 scale, but operators like the current scoring system. The CFMs emphasized that the current scoring system helps to avoid the potential for misusing test results as a promotion discriminator.

Additional communication was also identified as an important requirement for educating operators on a range of specific topics, including the grace period that will be used to allow operators to take the test, receive feedback, practice, and train. The CFMs emphasized that there is a misconception among some operators who think the implementation will be a "light switch"—one day, the tests and standards will be "turned on," and everyone will be held accountable.

Other BA CFMs

There were two primary concerns among other CFMs. First, CFMs questioned the time it would take to administer all of the tests. Specifically, some questioned whether it would be possible to administer the full test battery to an entire squadron in one day. Part of this concern may stem from the fact that the operator test is currently administered by pairing operators. The proposed test battery would likely require a higher evaluator-to-test taker ratio to ensure a sufficient number of test administrators were available to test everyone in one day.

Second, other comments raised concerns over the equipment needed to administer the test. For example, will it be possible to test operators in an austere environment, where gym equipment may not be readily available? Each piece of equipment required comes with a cost to purchase and maintain.

Concerns with the Specific Recommended Test Battery

TACP and ALO CFMs

There is some concern that commanders will go to extremes in prioritizing fitness. Fitness is clearly important to the career field, but CFMs do not want commanders to lose their focus on the mission. For example, CFMs suggested that an important objective would be to try to integrate physical movements in the "field," rather than overemphasizing gym time (e.g., perform lunges in the field with a rucksack instead of lunges in the gym). Overall, the CFMs indicated that commanders needed to balance the relative importance of fitness with other mission-essential tasks.

Other BA CFMs

CFMs indicated that the scientific process used to recommend the final ten tests in the proposed test battery (from the 39 possible tests included in the criterion-related validation study) was not well communicated or vetted among the career fields. There continue to be concerns with a total replacement of the current operator test rather than making more modest changes to address some of the recognized deficiencies. CFMs indicated that they understand the importance of improving and have shown this in the past by adopting better physical training exercises. CFMs further emphasized that any consideration of an updated test battery needs to address the utility of an update—that is, if operators have been successful on the battlefield so far, then what's the relative benefit of the new test battery to the career field? More communication is needed to explain why the current operator test battery is insufficient and to identify the problems/deficiencies that the new test battery fixes.

Overall, CFMs felt that this test battery has too many tests, will be difficult to implement and administer, and is difficult to communicate and manage among the many locations in which operators reside. For example, one CFM questioned, "Why not start with the six most critical tests up front and then perhaps you can phase in the other tests?"

In general, CFMs agreed that the current operator test battery can be improved and that there is value in improving the current system, but that additional communication and coordination with the career fields is needed to make sure the operational community receives what it needs and wants. Moving in this direction may require increased collaboration that includes the strength and conditioning coaches who work with BA operators.

CFM Concluding Thoughts

TACP and ALO CFMs

Previous communication about testing is often forgotten. Very few recall all of the steps in the process used to develop the test battery and standards, even among the participants in one or more phases of the previous studies. A timeline to document the history of the process would be helpful. It would need to be shared directly with everyone in the community, not just sent down through the chain of command.

Other BA CFMs

CFMs raised questions about who is the authority for driving the recommendation(s) and test implementation forward. Some indicated that the Operations (A3) component of the Air Staff should be in control of the process. Overall, CFMs acknowledged that the studies used to make recommendations have had a positive influence even though the test battery will be unlikely to be adopted in its current state.

6. Evaluation Framework

Based on the initial results of our evaluation, we present a framework that is designed to raise awareness and promote the systematic evaluation of a broader range of topics and concerns that may emerge as the Air Force moves forward with its implementation plan. Up to this point, we have been focusing on one element of this framework, which concentrates on individual-level attitudes and potential concerns.

In this chapter, we first describe the main objectives of the evaluation framework, followed by a discussion of possible evaluation priorities for the Air Force, and conclude with a summary of topics and concerns representative of the different categories and levels specified in the framework. A more comprehensive list of topics and concerns is provided in Appendix E.

Developing an Evaluation Framework

We developed an evaluation framework to help the Air Force assess O-PTB implementation and future tests designed for technical training students and recruits for TACPs, ALOs, and other BA specialties. This framework is intended to present a series of broader short- and long-term topics and research questions that can be used to guide evaluation efforts over time and can facilitate the ongoing monitoring of the O-PTB implementation across multiple levels. Many of the issues identified in the framework expand beyond our initial evaluation efforts presented in the previous chapters, and not all of them may be relevant to the Air Force's future evaluation efforts. Therefore, we recommend that the framework be reviewed to prioritize issues for evaluation. We also recommend that the evaluation framework be revisited periodically to ensure that high-priority issues are being evaluated and that emerging priorities are included in the framework. When specific issues emerge as priorities, the Air Force can use the evaluation framework to identify potential metrics and research methodologies that may be needed to address that issue.

We first describe inputs to the evaluation framework, which included interviews and working meetings with key stakeholders (e.g., BA CFMs), consultations with RAND experts in the fields of evaluation and military sociology, and a literature review. Next, we present a high-level summary of our evaluation framework to help raise awareness of potential issues during and following implementation of the O-PTB.[1] We emphasize that this framework is intended to be flexible, and additional issues of interest can be added over time to ensure that the framework captures topics of interest to Air Force stakeholders, such as CFMs and operators. Similarly, issues can be removed from the framework if they have no value for stakeholders.

---

[1] The full evaluation framework is provided in Appendix E.

Our initial evaluation efforts focused on topics relevant to the immediate priorities of the Air Force, but a broader perspective on evaluation is beneficial to identify potential concerns or issues as different priorities emerge over time (e.g., focus on tests and standards for recruits) and across different levels (e.g., individual, unit, career field). Consequently, our more-comprehensive framework identifies a range of possible topics that are loosely organized in a framework often used in the military for identifying requirements and potential gaps for a given set of strategic objectives.[2] A more-thorough discussion of this framework is provided in the following sections and Appendix E.

Broadly, the objectives of the evaluation framework are to (1) raise awareness of potential challenges and concerns for relevant stakeholders—those directly or indirectly impacted by implementation—during the implementation and adoption of the new physical tests and standards, and (2) promote the development of systematic data collection to monitor progress over time.

Sources Used to Develop the Evaluation Framework

We used three main sources to develop the evaluation framework: inputs from stakeholders, a review of literature emphasizing test-taker reactions, and consultation with internal RAND experts.

Elicitation of Stakeholder Input

Stakeholder input was generated through three main channels—conversations with the project sponsor, semistructured interviews with BA CFMs, and a working meeting with CFMs. First, regular meetings with the project sponsor throughout the period of performance yielded many issues of relevance to an evaluation of the implementation of the O-PTB. Second, in June 2016, we conducted semistructured interviews with seven CFMs about their perspectives on the PAST and plans for potentially adopting the O-PTB. These CFMs represented the following BA specialties: TACP, ALO, PJ, CRO, SOWT, and STO. Finally, in July 2016, we conducted a working meeting with some of the same CFMs, as well as additional CFMs, to gain a more in-depth understanding of issues relevant to the implementation of the O-PTB. In total, three CFMs attended, along with the chief of the AF-ESU, an Air Force statistician who was a member on the lead analytic team evaluating the criterion-related validity of different fitness tests, and three other SMEs from the TACP and ALO communities. The following BA specialties were represented at the working meeting: TACP, ALO, and CRO.

Review Literature on Test-Taker Reactions

As discussed in the previous chapters, understanding BA operators' reactions to the O-PTB was identified as a key topic of importance. To inform our understanding of issues related to this

---

[2] DOTMLPF-P is defined in Defense Acquisition University, 2017.

topic, we strategically reviewed several key articles on applicant and trainee reactions to assessments (see Appendix B). The literature reviewed was informative in identifying several important topics that may influence the successful implementation of an O-PTB. For example, operators may be less supportive of implementation if they believe the tests are unfair, irrelevant, or do not provide information on their job-related capabilities.

Consultation with RAND Subject-Matter Experts

We employed the expertise of two RAND colleagues who have conducted work relevant to this project. In particular, we spoke with senior political scientist Agnes Gereben Schaefer, who developed a monitoring framework for integrating women into the Marine Corps infantry (Schaefer et al., 2015). We also spoke with senior military sociologist Laura Miller to help identify issues relevant at the societal level (see explanation of levels below).

Organization of the Evaluation Framework

Although the framework could be structured in any number of ways, we opted to align it with the DOTMLPF-P framework (see Table 6.1 for definitions of each DOTMLPF-P category).[3] This structure is a familiar way to frame changes in military capability and has been useful for other military evaluation frameworks (Schaefer et al., 2015). Furthermore, this structure aligns with our objectives of raising awareness of potential challenges and concerns for relevant stakeholders and promoting the development of systematic data collection to monitor progress over time.

We added a category,[4] termed "Attitudinal," that encompasses attitudes and beliefs about the O-PTB and implementation process. Our initial evaluation efforts largely focused on attitudinal issues (e.g., reactions to tests), but other issues emerged in the open-ended comments and discussions with CFMs. For example, some CFMs raised questions about the time to administer the O-PTB, which would fall under "Organization" in this framework.

---

[3] AcqNotes, "JCIDS Process: JCIDS Manual of Operations," updated July 12, 2017.

[4] Each category is broadly divided into topic areas to better characterize specific evaluation issues. These topic areas are included as part of the evaluation framework presented in Appendix E.

Table 6.1. Category Definitions and Example Evaluation Issues

Category	Definition	Example Evaluation Issue
Organization	"The joint staffing (military, civilian, and contractor support) required to plan, operate, sustain, and reconstitute joint warfighting capabilities."	How does the time and cost of conducting the PAST differ from the time and costs required to implement the O-PTB?
Training	"Training, including mission rehearsals, of individuals, units, and staffs using joint doctrine or joint tactics, techniques, and procedures to prepare joint forces or joint staffs to respond to strategic, operational, or tactical requirements. . . . Training also pertains to nonmaterial aspects of operation and maintenance of materiel solutions."	How does the time needed to train for the O-PTB compare to the time needed prior to their implementation?
Materiel	"All items (including ships, tanks, self-propelled weapons, aircraft, etc., and related spares, repair parts, and support equipment, but excluding real property, installations, and utilities) necessary to equip, operate, maintain, and support joint military activities without distinction as to its application for administrative or combat purposes."	What financial costs are associated with storing and maintaining equipment needed for testing?
Leadership and Education	"A learning continuum that comprises training, experience, education, and self-improvement."	Are operators receiving the support and information they need to meet the O-PTB standards?
Personnel	"The personnel component primarily ensures that qualified personnel exist to support capability requirements across the joint force."	Is the O-PTB being administered and applied fairly to all BA in the career field?
Facilities	"Real property consisting of one or more of the following: buildings, structures, utility systems, associated roads and other pavements, and underlying land."	Do individuals have access to facilities in which to train for the O-PTB?
Policy	"Any [U.S. Department of Defense], other U.S. government agency/department, or international policy issues that may be changed to close or mitigate a capability gap, or if unchanged, prevent effective implementation of changes in the other seven DOTMLPF-P elemental areas."	Do policies exist regarding test administration and frequency? Are they followed?
Attitudinal	Encompasses attitudes and beliefs about the O-PTB and implementation process.	What are operator reactions to the O-PTB?

NOTE: The Doctrine category has been omitted from this table as it is not relevant to this evaluation. All definitions are drawn verbatim from the Joint Capabilities Integration and Development System manual, with the exception of the definition for the Attitudinal category, which was drafted by the report authors.

Building off this DOTMLPF-P-inspired structure, we identified other relevant considerations to assist in our framework's usefulness and application to include the important topics, evaluation issues, relevant metrics, recommended methods, and additional stakeholder guidance. This framework is not designed to be comprehensive, but rather to introduce a set of organizing principles that could be used to guide future evaluation efforts. To introduce the organization of our evaluation framework, Table 6.2 provides a subset of example questions posed at the individual level of analysis. A more complete framework for the individual level of analysis, as well as the frameworks for the other levels of analysis (e.g., career field, unit), are presented in Appendix E.

Table 6.2. Organization of Evaluation Framework: A Subset from the Individual Level

Category	Topic	Issue	Metric	Method	Stakeholder Guidance
Organization	Readiness	How does individual readiness compare with readiness prior to introduction of the O-PTB?	Comparison with metrics prior to implementation of the O-PTB	Primary data collection	May need to develop new data collection system
Training	Resources needed	How does the time needed to train for the new O-PTB tests compare to the time needed prior to their implementation?	Comparison with metrics prior to implementation of the O-PTB	Primary data collection	May need to develop new data collection system
Materiel	Access to equipment	Do individuals have access to test equipment on which to train?	Review of equipment available at Air Force installations with BA	Primary data collection	May need to develop new data collection system
Leadership and Education	Guidance for performance improvement	Is the guidance provided for improving performance on the O-PTB sufficient for improving performance?	Perceptions of guidance; comparison of performance data pre- and post-provision of guidance	Primary data collection; secondary data collection	Surveys, focus groups, and interviews could be conducted among BA operators, PTLs, strength and conditioning coaches, and training instructors
Personnel	Attrition	To what degree is failure to meet the O-PTB standards a reason for attrition?	Identify PF-related reasons for attrition	Primary data collection; secondary data collection	Surveys and interviews could be conducted among BA operators, CFMs
Facilities	Access to facilities	Do individuals have access to facilities in which to train for the PF tests?	Review of facilities available at Air Force installations with BA	Primary data collection	May need to develop new data collection system
Policy	Policy	What policies regarding the O-PTB are in place for operators without a unit?	Identification of policies for operators without a unit	Policy review	
Attitudinal	Test perceptions	What is the acceptability of the new O-PTB among BA operators?	Identification of acceptability	Primary data collection	Interviews, focus groups, surveys could be conducted among BA operators, PTLs, CFMs

Important Topics

For each of the categories, there may be a number of relevant topics that add specificity to further organize potential issues. For example, although not shown in Table 6.1, for personnel, we identified six possible topics: entry, retention, promotion, attrition, injury, and resources needed. Identifying various topics within the personnel category encourages more comprehensive coverage of the category domain.

Evaluation Issues

The framework includes corresponding issues for each topic. For example, one key Leadership and Education evaluation issue relevant to the individual level (framed as an evaluation question) focuses on performance improvement: "Is the guidance provided for improving performance on the O-PTB sufficient for improving performance?" Other example evaluation issues corresponding with each category are included in Table 6.1.

Metrics

The "metrics" column describes the information needed to address the evaluation issue. This column provides a brief description of the data comparison or characterization that would need to be made. Using our previous example of an evaluation issue concerning the Leadership and Education category at the individual level (i.e., "Is the guidance provided for improving performance on the O-PTB sufficient for improving performance?"), the metrics we recommend for addressing this issue include (1) perceptions of guidance and (2) performance data comparing pre- and postprovision of guidance.

Recommended Methods and Stakeholder Guidance

In addition to recommending metrics for assessing each evaluation issue, we recommend methods for collecting necessary metrics. Data-collection methods are broken into methods requiring primary or secondary data collection.

Primary data collection involves gathering new data to address evaluation questions and may require developing new data-collection processes and systems. It may not be feasible to stand up all the new data-collection processes and systems that we suggest, but we present them to help guide decisionmaking concerning evaluation of O-PTB implementation.

Common methods of primary data collection to gather information about individuals' knowledge, attitudes, and perceptions of their experiences include surveys, focus groups, and interviews. Surveys allow for collection of data across a large number of respondents. Focus groups are useful for eliciting the perspectives of small groups of individuals and supporting in-depth discussion across members of the group. Interviews are typically conducted one-on-one and provide insight into individual experiences; therefore, they can be useful in eliciting detailed insight on sensitive topics. When we recommend the use of surveys, focus groups, or interviews, we provide guidance on the different stakeholder groups from which data should be collected in the stakeholder guidance column of the framework.

Secondary data collection involves gathering existing data to address evaluation questions. Secondary data collection might include gathering data on individuals' performance on the tests relative to standards; training performance data (e.g., field training exercises, full mission profiles); after-action reports; and medical records that identify injuries. Another form of secondary data collection is gathering policies or materials for further review and analysis.

We revisit our example evaluation issue concerning the Leadership and Education category at the individual level (i.e., "Is the guidance provided for improving performance on the O-PTB sufficient for improving performance?"). We recommend the following metrics: surveys; focus groups; and interviews with BA operators, PTLs, strength and conditioning coaches, and training instructors (i.e., primary data collection).

The Evaluation Framework

The evaluation framework is presented as a series of five tables (see Appendix E)—one for each level of the framework (i.e., individual, unit, career field, institutional, and societal). For each table (with the exception of the societal table), each row represents a single evaluation issue, phrased in terms of a question. For each issue, the DOTMLPF-P category and topic are identified, as well as metrics for assessing the evaluation issue. Suggested methods appear in the "Method" column, and when surveys, focus groups, or interviews are recommended, the stakeholders from whom such data should be collected appears in the "Stakeholder Guidance" column.

The societal level table (see Table E.5 in Appendix E) is presented as a list of evaluation issues, without accompanying information about category, topic, metrics, or methods. This is because the societal level table is broad in scope and does not clearly align with the DOTMLPF-P framework. We note that when evaluation issues relate to differences between men and women, the purpose is not to encourage differential standards within the Air Force for male and female airmen. Rather, the purpose is to identify issues that may be raised (by Congress, members of society, current airmen, and veterans of the Air Force) in relation to changing to the O-PTB.

7. Conclusion and Recommendations

This project focused primarily on one of the main IVT questions: various concerns of key stakeholders related to the O-PTB. We conducted a preliminary evaluation of the perceptions of stakeholders in the first wave of the Air Force's implementation of the physical tests and standards for TACPs and ALOs. We began our efforts by determining relevant stakeholders, which included TACP and ALO operators, CFMs, and PTLs. Then, we identified topics relevant to each stakeholder group. Specifically, we developed evaluation surveys that could be used to gather feedback from test takers and test administrators. We also held meetings with CFMs of BA specialties to identify their perspectives on what factors concern them and issues that need to be addressed to facilitate the implementation of new tests and standards. Understanding the perspectives of these key stakeholders is an important step in identifying potential barriers to implementation and uncovering opportunities for further strengthening the operational physical test battery, test procedures, and training of test administrators.

The evaluation surveys we developed were used during the implementation period to solicit initial feedback from TACPs and ALOs and PTLs on the prototype operator physical test battery. TACPs and ALOs responded to four primary statements about each test:

1. The [O-PTB subtest] was administered to all operators in the same way.
2. I know how I performed on the [O-PTB subtest] relative to the required standard.
3. I am concerned that I was injured or could have been injured while completing the [O-PTB subtest].
4. Taking the [O-PTB subtest] was frustrating.

The evaluation surveys also measured test taker and test administrator global evaluations of the full test battery. In addition to collection evaluation information from operators, we met with CFMs to better understand their perspective on their primary concerns regarding test implementation. In Table 7.1, we summarize the primary conclusions regarding the feedback from TACP and ALO operators, PTLs, and CFMs, and present a recommendation to address each of the primary findings.

Table 7.1. Summary of Findings and Recommendations

Finding	Accompanying Recommendation
1. Overall, TACPs and ALOs indicated consistently strong, positive support for the O-PTB.	Communicate results broadly throughout the TACP and ALO community; disseminate results to other BA leaders.
2. TACPs and ALOs generally felt that each test was administered to all operators in the same way and that they knew how they performed relative to the standard.	Test administrators were being observed and had just received training on how to administer the test; therefore, follow-up evaluations should be conducted to ensure consistent administration continues to be followed.
3. TACPs and ALOs indicated concern about the potential for injury for the Trap Bar Deadlift (20 percent), and PTLs expressed concern that operators could be injured while taking the test battery (12 percent).	Consider (1) further training on proper form and technique, (2) increasing the opportunities to practice the tests and receive feedback, and/or (3) modifying test administration instructions.
4. TACPs and ALOs were most frustrated by the Extended Cross Knee Crunch (41 percent) and the Medicine Ball Toss (21 percent), and 15 percent of PTLs indicated that operators seemed frustrated by the test.	
5. TACPs and ALOs (18 percent) and PTLs (12 percent) did not feel that the test battery would be fair for all TACPs regardless of rank, age, stature, gender, or race/ethnicity. TACP and ALO CFMs echoed this sentiment by expressing concern about the community's lack of awareness regarding the scientific validation process to determine the tests included in the battery.	Deliver additional communication about the history of the test-development process, how tests were selected, and how they link to job and mission-related requirements. Consider relevance of tests and standards to TACPs/ALOs across organizational levels (e.g., staff position).
6. TACPs and ALOs (71 percent) and PTLs (78 percent) felt that it was important that test administrators be other TACPs; in contrast, CFMs emphasized that the test administrator could be anyone.	Consider the advantages and disadvantages of various test administrator characteristics.
7. Other BA CFMs (i.e., not TACP or ALO) expressed strong concerns over logistical issues regarding test administration (e.g., time, equipment cost).	Examine the time required, on average, to administer the prototype test battery and the cost to purchase all equipment for a squadron of 100.
8. Other BA CFMs (i.e., not TACP or ALO) recognized deficiencies in the PAST and expressed interest in addressing these shortcomings through a more collaborative effort.	Increase frequency of communication among CFMs, commanders, strength and conditioning coaches, and the AF-ESU. Consider trade-offs between scientific validity and other career field needs including feasibility, cost, and perceived utility.

Overall, the results from TACPs, ALOs, and PTLs were very positive, with relatively few concerns identified among a minority of TACP and ALO operators who participated in this phase of test implementation. Most indicated that each test was administered correctly and that the test battery measured important job-related physical abilities. Many open-ended comments further supported these findings with positive comparisons to the PAST, which stated that the O-PTB is more comprehensive and more representative of operational tasks. Only a few tests caused frustration or raised concerns for potential injury among some of the respondents. Three tests in particular warrant close monitoring and further review to identify opportunities to address potential concerns for injury and frustration: the Extended Cross Knee Crunch, Trap Bar Deadlift, and Medicine Ball Toss. For each of these tests, we recommend additional training, practice, and feedback on the proper form and technique, which should help address these concerns. Although operators did not report any new injuries from training for or completing these tests, future evaluation efforts should document injuries that may occur as a result of testing or test preparation. Additional data collection may also be required to fully evaluate the utility of the Extended Cross Knee Crunch because this was a new test specifically designed as part of the Air Force validation study. Therefore, we recommend comparing the Extended Cross Knee Crunch test against other tests designed to measure core strength and endurance.

Analysis of open-ended comments in combination with CFM feedback suggests that additional structured communication could help address a range of concerns (e.g., career repercussions if one fails the O-PTB) and further educate the operators on the purpose of the test battery and the steps that were taken inform the decision to select the recommended tests and standards. Consequently, we recommend developing short, informative pamphlets that can be used to answer frequently asked questions. For example, operators could benefit from additional education on the role of each test in assessing their physical capability to perform CPTs of their specialty. Operators could also benefit by having additional information on the implementation plan, timeline, and potential changes that may be made to the protocols for specific tests (e.g., Farmer's Carry).

Although the results were very positive, our analysis represents only a snapshot of perceptions at a given point in time. As the implementation proceeds for TACP and ALO operators, we recommend following up with future evaluations to determine if perceptions have changed. Future evaluations should be conducted annually for the first two years of implementation and then every three to five years thereafter. Tracking individual operators over time using pre- and post-test research designs will allow for more sophisticated analyses for evaluating the effectiveness of different interventions (e.g., additional training on form/technique) to address concerns.

In addition to addressing the immediate priorities of the Air Force in regard to implementing tests and standards for the TACP/ALO specialties, we also met with other BA CFMs to identify their concerns with the recommended O-PTB. The majority of their concerns reflect three substantive issues: First, other BA CFMs indicated that there was insufficient information provided on how the AF-ESU reduced the number of tests in the validation study from 39 to the final recommended test battery consisting of 10 tests. Second, they communicated concerns regarding the high number of tests included in the final battery and that managing this many tests would be difficult both in terms of administration and in terms of equipment resourcing and maintenance. Finally, the other BA CFMs did not see an urgent need to update their tests and standards. The TACP/ALO specialties do not currently have an existing operator test, whereas, the other BA specialties have had an operator test that has been in place for many years. In general, the consensus among other BA CFMs was that improvements could be made to the existing operator test but that a complete overhaul is not required.

Taking into account the broader scientific literature on organizational change, we strongly recommend additional discussions to determine what, if any, value can be gained by updating the existing operator test for the other BA specialties. It is unlikely that any changes will occur unless there is commitment from career field leaders and commanders in the operational community. On the whole, the other BA CFMs recognize that the existing operator test is deficient and is not as comprehensive as it could be in measuring important physical ability required to perform critical physical job and mission tasks. However, there is a clear desire among other BA CFMs to have more control in the process of making the final selections for an updated operator test battery. Although grandfathering in current operators to potential new requirements may ease some stakeholder concerns, this approach could result in negative perceptions of the readiness of more senior operators. Nonetheless, implementing new requirements gradually over time can provide sufficient opportunities to fine-tune the tests, training guidelines, and test administration. Related readiness concerns can be further addressed by allowing for a sufficient period of adaptation (e.g., one to two years) to the new requirements prior to taking administrative actions for failing to meet the new requirements.

The valuable insights garnered from our preliminary evaluation efforts helped to inform our framework development, which is intended to monitor and evaluate the implementation of an occupationally specific PF test battery for the TACPs and ALO communities. This evaluation framework should be used to raise awareness of possible issues that may influence the successful implementation of PF tests and standards and to guide evaluation efforts for determining how well implementation objectives are being met. Although the objective of this report was to provide feedback specifically on the implementation of the O-PTB, the framework provided can be extended to address future potential implementation of an updated test battery for recruits and trainees. Many of the issues will remain relevant, whereas other issues will be unique to the recruiting and/or training context.

Appendix A. Description of O-PTB Components

The O-PTB descriptions were developed by the Air Force. For each test, there is a stated purpose, detailed protocol, and, if relevant, necessary equipment.

Grip Strength

PURPOSE: To measure muscular strength of the lower arm.
EQUIPMENT: Hand dynamometer, chalk.
PROTOCOL:

1. Before using the handgrip dynamometer, adjust the handgrip size to a position that is comfortable for you.
2. Stand erect, with your arm and forearm positioned as follows: shoulder down and to the side, elbow flexed at 90°, forearm in the neutral position, and wrist in slight extension (0° to 30°).
3. You will squeeze the dynamometer as hard as possible using one brief maximal contraction and no extraneous body movement.
4. I will administer two trials for each hand, allowing a 1-minute rest between trials, and use the best score as your static strength.

Medicine Ball Toss (Backwards, Sidearm, Overhead)

PURPOSE: To determine upper extremity strength and power.

Backwards Medicine Ball Toss

EQUIPMENT: 20-lb medicine ball, tape measure, marking chalk.
PROTOCOL:

1. Stand with your heels on the starting line.
2. Hold the medicine ball with both hands at hip level.
3. Bend at the knees to gain momentum and toss the medicine ball from an underhand position over your head.

Sidearm Medicine Ball Toss

EQUIPMENT: 20-lb medicine ball, tape measure, marking chalk.
PROTOCOL:

1. Grasp the medicine ball with both hands.
2. Align both feet in a shoulder-width stance parallel to the line.
3. Rotate your trunk in the direction opposite to the throwing direction as a counter-movement.

4. Follow this motion by rotating your trunk in the throwing direction as you throw the medicine ball as far as possible.

Overhead Medicine Ball Toss

EQUIPMENT: 8-lb medicine ball, tape measure, marking chalk.
PROTOCOL:

1. Stand facing the starting line.
2. Hold the medicine ball with both hands overhead.
3. Bend at the knees to gain momentum, bring the medicine ball behind the neck and toss the medicine ball from an overhead position.

ADMINISTRATIVE NOTE: Allow one familiarization toss and two trials for scoring for all medicine ball tosses, record all distances thrown. Measurement is made from the front of the line to the point where the ball lands (center of the ball).

Three-Cone Drill Test

PURPOSE: To test multidirectional speed, agility, and body control.
EQUIPMENT: Three cones, nonslip testing surface, standard stopwatch, or electronic testing device.
PROTOCOL:

1. There are three cones in an upside-down "L" configuration. Each cone is placed 5 yards (4.6 meters) apart from the others.
2. You will start from the prone position and sprint from the starting line at cone A to cone B, touching the line at cone B.
3. Then reverse direction and sprint back to cone A, touching the line at cone A.
4. Change direction and sprint around cone B to cone C.
5. Circle cone C keeping cone C to your left and sprint back around cone B and past the original starting line at cone A.
6. You will complete three trials of this test with a minimum of 1-minute rest between each trial.

ADMINISTRATIVE NOTE: Test subject must touch the line at cone B and the line at cone A. Mark out the slowest of the test subject's three trials.

5 RM Trap Bar Deadlift

PURPOSE: Measures anaerobic lower extremity power.
EQUIPMENT: Sorinex Diamond Bar, bumper plates.
PROTOCOL:

1. Stand with feet parallel in trap bar, weight to each side.
2. Place the hands on the trap bar slightly wider than shoulder-width apart, outside of the knees, with the elbows fully extended.

3. Place the feet flat on the floor and position the subject within the trap bar.
4. Position the body with the back flat or slightly arched, trapezius relaxed and slightly stretched, chest held up and out, head in line with the vertebral column or slightly hyperextended, heels in contact with the floor, shoulders over or slightly in front of the bar, and eyes focused straight ahead or slightly upward.
5. All repetitions will begin from this position.

   a. Upward movement phase: Lift the bar off the floor by extending the hips and knees. Keep the torso-to-floor angle constant; do not let the hips rise before the shoulders. Maintain a flat-back position. Continue to extend the hips and knees until the body reaches a fully erect torso position.
   b. Downward movement phase: Allow the hips and knees to flex to slowly lower the bar to the floor. Maintain the flat-back body position; do not flex the torso forward.

Pull-Up Test

PURPOSE: The pull-up test is an endurance test used to assess the forearm and upper arm flexor muscles.

EQUIPMENT: Standard pull-up bar, stopwatch.

PROTOCOL: 2-minute exercise.

1. Pull-ups are a two-count exercise. The starting position is you hanging from a bar, your palms facing away from you with no bend in your elbows ("dead hang"). Your hands are spread at approximately shoulder width apart.
2. Count one; pull your body up until your chin is over the bar.
3. Count two; you will then return to the starting position.

NOTE: Legs are allowed to bend, but must not be kicked or manipulated to aid upward movement. If you fall off or release your hands from the bar, the exercise will be terminated. The number of repetitions properly completed in 2 minutes will be recorded as your score.

ADMINISTRATIVE NOTE: Please administer this test to two test subjects at a time.

Lunges, 50-Lb Sandbag

PURPOSE: To determine muscular strength and endurance of the core and lower body.

EQUIPMENT: 50-lb sandbag, metronome (56 beats per minute [BPM]).

PROTOCOL:

1. You will start in a standing position with a 50-lb sandbag resting behind your neck and across your shoulders.
2. This test is carried out by stepping forward, kneeling down, and with the same leg pushing your body upright and stepping back.
3. You will alternate right and left leg, so that the heel of the front leg is a minimum 10 cm in front of the knee of the back leg and the knee of the back leg must reach a distance of at least 10 cm from the floor.
4. Your back must remain stable and your knee should not pass your toes while lunging.

5. The lunge is repeated, in cadence (56 BPM), to exhaustion.
6. If the repetitions are not in cadence with the metronome for two or more repetitions the test will be terminated. The goal of this event is to complete as many correct repetitions as possible.
7. Report your heart rate and rate of perceived exertion (RPE) to the test administrator immediately upon completion of the test.

ADMINISTRATIVE NOTE: Do not announce the maximum or minimum reps. Stop the subject at 5 minutes.

Extended Cross Knee Crunch

PURPOSE: The cross knee crunch is used to assess a test subject's muscular fitness.
EQUIPMENT: Level ground or mat, stopwatch, 56-BPM metronome.
PROTOCOL:

1. Lie on your back and place your feet on the ground with your legs at 90 degrees.
2. Cross your arms over your chest and lock your hands into your armpits.
3. Curl up and bring your left elbow and shoulder across your body and reach with your left elbow to touch your right knee.
4. In a controlled movement, return to lying flat on your back.
5. Next, curl up and bring your right elbow and shoulder across your body and reach with your right elbow to touch your left knee.
6. In a controlled movement, return to lying flat on your back.
7. Each knee will count as one repetition. Completed repetitions will be counted in the down position.
8. Your score will be recorded as the number of complete repetitions you complete to a 56-BPM metronome.

Farmer's Carry

PURPOSE: Measures lower body, core, forearm, and grip strength.
EQUIPMENT: Two 50-lb sandbags with handles, four cones, stopwatch.
PROTOCOL: Place two cones at the start and finish.

1. Stand between two sandbags positioned behind the starting line. You may lift and put down the sandbags to familiarize yourself with the weight.
2. You will squat down between the two 50-lb sandbags and grasp the handles.
3. On the command "Go" the timer will start the clock and you will stand and walk/jog/run for 100 yards.
4. I will record your time from the command "GO!" until you cross the finish line.
5. If you drop the sandbag(s), you may pick up the sandbag(s) and continue to the finish line.
6. If you drop the sandbag(s) and fail to complete the 100-yard distance, the distance and the time you completed will be recorded.

ADMINISTRATIVE NOTE: Start with two groups on the opposite ends of the track.

1,000-Meter Ergometer Row Test

PURPOSE: Measure total body anaerobic capacity.

EQUIPMENT: Concept IID rowing ergometer (or equivalent) and stopwatch.

PROTOCOL:

1. The Concept IID will be set for 1,000 meters and the drag factor will be set to five.
2. Sit ready to start the 1,000-meter test.
3. The aim of the test is to cover the 1,000-meter in the shortest possible time.
4. Your time will start on your first pull.
5. The row machine will keep track of your time.
6. Report your heart rate and RPE to the test administrator immediately upon completion of the test.

1.5-Mile Run Test

PURPOSE: This 1.5-mile timed run is used to measure cardio-respiratory fitness.

EQUIPMENT: Establish a standard course of accurate distance that is as level and even as possible, stopwatch.

PROTOCOL:

1. Prior to beginning the 1.5-mile run, you may complete up to a 3-minute warm up.
2. You will line up behind the starting line and will be instructed to begin running as I start the stopwatch.
3. No physical assistance from anyone or anything is permitted.
4. You are required to stay on and complete the entire marked course. Leaving the course is disqualifying and terminates the test.
5. Your completion time will be recorded when you cross the finish line, and you are required to complete a cool down for approximately 5 minutes.
6. If at any time you are feeling in poor health, you are to stop running immediately and you will be given assistance.

Report your heart rate and RPE to the test administrator immediately upon completion of the test.

Appendix B. Literature Review of Applicant and Trainee Reactions and Item Mapping

To inform our potential pool of topics for the evaluation instruments, we began by reviewing two literatures that have emerged from relatively distinct approaches: applicant and trainee reactions. These reactions refer to subjective experiences of the hiring process or training experience, respectively (e.g., Kirkpatrick, 1996; Ryan and Ployhart, 2000). This appendix briefly presents seminal theories, influential classification schemes, and widely used measures for each reaction type.

Applicant Reactions

Traditionally, research on the hiring process has predominately adopted the organization's perspective; however, applicant (or job candidate) reactions have recently received greater research and practitioner attention because of their economic, legal, and psychological implications (Hülsheger and Anderson, 2009). For this study, we are particularly interested in how applicant reactions may influence attitudes and/or behavior (i.e., test and job performance).

Theoretical Models

Various theoretical models have emerged and can be summarized by three lines of research: (1) test perceptions, (2) organizational justice, and (3) attributions (Ployhart and Harold, 2004). Test perceptions research represents early work on applicant reactions by examining how applicants' perceptions of a test influence their performance on that test and therefore the inferences that can be drawn from a test. Importantly, researchers sought to examine explanations for test performance differences such as motivation and concentration in addition to ability-related factors (Arvey et al., 1990). In line with this model's predictions, subsequent studies have demonstrated some support for the importance of test attitudes and motivation for test performance (e.g., Chan et al., 1998; McCarthy et al., 2013; Schmit and Ryan, 1992).

Arguably, organizational justice theories (e.g., Greenberg, 1990), which emphasize procedural and distributive justice rules, have exerted the most influence on applicant reactions research (for a recent review, see Colquitt and Zipay, 2015). Procedural justice focuses on the perceived fairness of how decisions get made (i.e., process), whereas distributive justice focuses on the perceived fairness of the decision itself (i.e., outcomes). Gilliland (1993) extended these concepts to a selection context by proposing that applicants' perceptions of the fairness of the selection process will influence their attitudes (e.g., toward the organization), intentions (e.g., to accept a job offer), self-perceptions (e.g., self-confidence), and behaviors (e.g., test and job performance). Ryan and Ployhart (2000) expanded on Gilliland's model to include, among other revisions, additional antecedents (e.g., individual characteristics such as work experience and personality) as well as additional moderators (e.g., job desirability and selection ratio). Hausknecht, Day, and Thomas (2004) revised this model yet again by further elaborating on the antecedents and outcomes. In line with its predictions, meta-analyses have demonstrated support for many aspects of this model (Hausknecht, Day, and Thomas, 2004; Truxillo et al., 2009).

Attribution theories focus on understanding the process of how individuals decide the cause(s) driving the behavior of themselves and others (for a review, see Fiske and Taylor, 1991). This line of research differs from organizational justice because it considers applicant reactions to be a result of individuals' judgment of causality instead of individuals' perceptions of justice. This, in turn, shifts the focus from understanding antecedents and consequences of justice perceptions to individuals' understanding of reasons causing the behaviors. Indeed, there is a broad literature in social psychology examining attributions, such as the fundamental attribution error—that is, the tendency to attribute an individual's behavior to dispositional factors ("he's a mean person") rather than situational factors ("he's having a bad day")—and the availability heuristic (the tendency to focus on information that is readily available). More recently, Ployhart and Harold (2004) proposed the Applicant Attribute–Reaction Theory, which leverages attribution research to predict behavioral consequences (e.g., test performance, withdrawal). Although this model is promising, it has yet to be evaluated empirically.

Taken together, there is considerable support for the idea that applicant reactions influence various issues, such as attitudes toward the organization, intent to accept a job offer, levels of self-confidence, and performance (both on the test and on the job). In the next section, we discuss how applicant reactions have been conceptualized by outlining relevant classification schemes (or taxonomies) and associated measures in Table B.1.

Classification Schemes and Existing Measures

Anchored in the organizational justice perspective, we outline two widely cited classification schemes: the Justice Model of Applicant Reactions (Gilliland, 1993) and the Social Validity Theory (Schuler, 1993). We introduce their corresponding measures: the Selection Procedural Justice Scale (Bauer et al., 2001) and the Social Process Questionnaire on Selection (Derous, Born, and De Witte, 2004). By *corresponding*, we mean that the model informed the development of the scale. For example, the Justice Model of Applicant Reactions guided the development of the Selection Procedural Justice Scale. The final measure is the Test Attitude Survey (Arvey et al., 1990), which emerged from the test perception perspective.

The purpose of this mapping is to highlight potentially relevant topics based on a review of the literature. For our final evaluation instrument, we included the majority of the topics from the Selection Procedural Justice Scale and Social Process Questionnaire on Selection as well as a few from the Test Attitude Survey.

Table B.1. Comparison of Selected Applicant Reactions Classification Schemes

Classification Schemes		Measures		
Justice Model of Applicant Reactions	**Social Validity Theory**	**Selection Procedural Justice Scale**	**Social Process Questionnaire on Selection**	**Test Attitude Survey**
Job relatedness (FC)		Job relatedness—content[a]	Objectivity	Belief in tests
		Job relatedness—predictive (St)		
Opportunity to perform (FC)	Participation[b]	Chance to perform (St)	Participation[b]	
Reconsideration opportunity (FC)		Reconsideration opportunity (St)		
Consistency (FC)		Consistency (So)		
Feedback (E)	Feedback	Feedback (St)	Feedback	
Selection information (E)	Informativeness	Information known (St)	Job Information	
Honesty (E)	Transparency	Openness (So)	Transparency	
Interpersonal effectiveness (IT)		Treatment (So)	Human Treatment	
Two-way communication (IT)		Two-way communication (So)		
Propriety of questions (IT)		Propriety of questions (So)		
				Motivation
				Preparation
				Test ease
				External attribution
				General need achievement
				Future effects
				Lack of concentration
				Comparative anxiety

NOTE: The letters in parentheses indicate the superordinate category that the "topics" (or dimensions) fall under. FC = formal characteristics; E = explanations; IT = interpersonal treatment; St = structure; So = social.
[a] Unlike the other "topics" (or dimensions), job relatedness–content is not associated with a superordinate category.
[b] In the Social Validity Theory (Schuler, 1993), participation refers to (1) involvement in deciding the specific selection situation, instrument, or process and (2) the general notion that individuals can maintain control over the situation or their behavior. When aligning participation from the Social Validity Theory with the opportunity to perform from the Justice Model of Applicant Reactions, we focus on the latter component of participation (i.e., perceived individual control).

Trainee Reactions

Literature on trainee reactions, which are commonly used to evaluate the effectiveness of training programs, is also directly relevant because we assess the implementation of PTL training. Although there has been much scholarly debate about what type of information that trainee reactions can convey; they have—unquestionably—been a hallmark to the training evaluation literature.

Theoretical Model

Focusing solely on the impact of training reactions, Kirkpatrick's model of evaluation (1976) is the most widely cited theoretical model of training outcomes. According to this approach, Kirkpatrick proposes four levels:

- Level 1: trainees' reactions (how well trainees liked training)
- Level 2: learning (knowledge, skills, etc.)
- Level 3: behavior (on-the-job performance)
- Level 4: results (objective training outcomes such as costs).

This approach assumes causality across levels such that positive reactions (Level 1) lead to learning (Level 2), which leads to behavioral change (Level 3), etc. By extension, this model suggests that higher levels are not expected to change unless appropriate changes occur at the previous level. Although this approach is intuitively appealing, criticism has been aimed at both its causality assumptions (e.g., Alliger and Janak, 1989) as well as the amount of support provided by empirical research (Sitzmann et al., 2008). When accounting for pretraining knowledge, training reactions were associated—to varying degrees—with cognitive and affective learning (Sitzmann et al., 2008). Specifically, training reactions were weakly related to post-training knowledge (both declarative and procedural), moderately related to post-training self-efficacy, and strongly related to post-training motivation.

Classification Schemes and Existing Measures

Classification schemes range from one overall facet to many facets. For instance, Morgan and Casper (2000) identified six facets: satisfaction with instructor, overall satisfaction, satisfaction with testing, utility of training, satisfaction with materials, and satisfaction with course structure. Others have identified two facets: affective reactions (i.e., the enjoyableness of training) and utility judgments (i.e., the perceived value of training) (e.g., Alliger et al., 1997; Tracey et al., 2001; Mathieu, Tannenbaum, and Salas, 1992). There have been various items used to assess these two facets of training. Examples for affective reactions include "I am pleased I attended this training" and "I enjoyed this training." Examples for utility judgments include "The training was relevant to my job" and "This training provided useful examples." In addition to these two facets, Warr and Bunce (1995) also added perceived difficulty (e.g., the effort required to perform well in training).

To reconcile these various classifications, Brown (2005) suggested training reactions may be conceptualized as one global satisfaction concept underlying reactions to specific aspects of training (also see Sitzmann et al., 2008). However, the purpose of evaluation is critical to determining the appropriate number of facets (Salas et al., 2012). For instance, an overall satisfaction measure may be useful for detecting general problems with training, whereas facet measures of satisfaction may be useful for diagnosing problems with specific elements of training.

Applying the Literature to Battlefield Airmen

To ensure that our evaluation instrument was relevant for our purpose, we sought to integrate specific concerns from the BA community within the context of evaluation (i.e., early stages of O-PTB implementation). In collaboration with the project sponsors (Vice Commander in Air Education and Training Command and Air Force Director of Military Force Management Policy, Deputy Chief of Staff for Manpower, Personnel and Services), we identified the following topics unique to our purpose:

- potential for injury
- desired characteristics of test administrators
- recommendations for improvement.

We began to assign topics appropriate to a given stakeholder based on information we gathered from the literature and stakeholders. Given the stakeholder, we then selected the most suitable method of data collection (e.g., evaluation survey for ALO and TACP operators and PTLs, semistructured interviews for CFMs) and finally developed items accordingly. The topics, description of topics, and representative questions are provided in Table B.2.

Table B.2. Items for Relevant Stakeholders

		Stakeholder Questions		
Topic	Topic Description	ALO and TACP Operators (Survey)	PTLs (Survey)	CFMs (Semistructured interviews)
Chance to perform/ fairness of tests	"Having adequate opportunity to demonstrate one's knowledge, skills and abilities in the testing situation" (Bauer et al., 2012, p. 18)	I could really show my physical abilities through this test battery. The testing process I just completed would be fair for all TACPs regardless of rank, age, stature, gender, or race/ethnicity.	Operators could really show their physical abilities through this test battery. The testing process would be fair for all TACPs regardless of rank, age, stature, gender, or race/ethnicity.	What concerns/ benefits, if any, do you have about the O-PTB [regarding chance to perform/ specific groups]?
Comparison to PAST	Directly comparing the current PF standards to the previous standards	This test battery is a better measure of my operational capabilities than the PAST. [Follow-up: Please tell us more about why you answered the previous question the way you did.]	This test battery is a better measure of operational capabilities than the PAST. [Follow-up: Please tell us more about why you answered the previous question the way you did.]	Do you see any benefits/challenges compared to PAST?
Job-relatedness	"Extent to which a test either appears to measure the content of the job or appears to be a valid predictor of job performance" (Bauer et al., 2012, p. 18)	Doing well on this test battery means a person can perform the physical job duties of a TACP well. This test battery clearly measures abilities required of a TACP.	Doing well on this test battery means a person can perform the physical job duties of a TACP well. This test battery clearly measures abilities required of a TACP.	What concerns/benefits, if any, do you have about the O-PTB [regarding job-relatedness]?
Concern for injury	The degree to which individuals are concerned about potential injury	I am concerned that I was injured or could have been injured while completing the [specific PF test].	I am concerned that TACPs may have been injured or could have been injured while completing this test battery.	What concerns/benefits, if any, do you have about the O-PTB [regarding potential for injury]?
Cause for frustration	The extent to which individuals subjectively evaluate their test-taking experience as negative (e.g., Alliger et al., 1997)	Taking the [specific PF test] was frustrating.	It appeared that taking this test battery was frustrating for TACPs.	What concerns/benefits, if any, do you have about the O-PTB [regarding frustration]?
Test administrator characteristics	Examine desirable characteristics of test administrators	It is important to me that the people who administer PF tests are other TACPs.	It is important that the people who administer PF tests are other TACPs.	What concerns/benefits if any, do you have about the O-PTB [regarding test administration characteristics]?

60

Topic	Topic Description	ALO and TACP Operators (Survey)	PTLs (Survey)	CFMs (Semistructured interviews)
Perceived utility of results	The extent to which individuals perceive testing results as useful (e.g., Alliger et al., 1997)	Knowing how well I did on the PF tests will help me improve my job-related physical capability. Other TACPs will find their test results useful for improving their own job-related physical capability.	TACPs will find their test results useful for improving their own job-related PF.	What concerns/benefits if any, do you have about the O-PTB [regarding utility of results]?
Concerns	Examine the potential for any possible concerns	I have concerns about this test battery. [Follow-up: Please tell us more about your concerns. [NOTE: Each test was listed separately with space for test-specific comments.]	I have concerns about this test battery. [Follow-up: Please tell us more about your concerns.] [NOTE: Each test was listed separately with space for test-specific comments.]	What other concerns, if any, do you have about the O-PTB?
Recommendations	Examine the potential recommendations for improvement	Please tell us any changes that you would recommend to the testing.	Please tell us any changes that you would recommend to the testing.	
Consistency of test administration	"Uniformity of content across test settings, in scoring, and in the interpretation of scores. Assurance that decision-making procedures are consistent across people and over time" (Bauer et al., 2012, p. 18–19)	[Specific PF test] was administered to all operators in the same way.	Each test was administered to all operators in the same way.	
Knowledge of performance	Extent to which performance on a test is readily apparent	I know how I performed on the [specific PF test] relative to the required standard.		
Information known	"Information, communication, and explanation about the selection process prior to testing" (Bauer et al., 2001, p. 391)	I understood in advance what tests I would be taking. I have been provided with information about how my test results will be used. I know the consequences of not meeting test standards.		

		Stakeholder Questions		
Topic	Topic Description	ALO and TACP Operators (Survey)	PTLs (Survey)	CFMs (Semistructured interviews)
Two-way communication	"The interpersonal interaction between candidate and test administrator that allows candidates the opportunity to give their views or have their views considered in the selection process" (Bauer et al., 2012, p. 19)	I was able to ask questions about the tests before and during the tests.		
Openness	"The importance of honesty and truthfulness when communicating with candidates, and in particular, in instances when either candidness or deception would likely be particularly salient in the selection procedure" (Bauer et al., 2012, p. 19)	Test administrators answered questions in a straightforward manner.		
Respectful treatment of candidates	"The degree to which candidates feel they are treated with warmth and respect by the test administrator" (Bauer et al., 2012, p. 19)	The test administrators treated me with respect during the testing process.		
Feedback	"Providing candidates with informative and timely feedback on aspects of the decision making process" (Bauer et al., 2012, p. 19)	I am satisfied with the feedback I received about my test results.		
Preparation	The amount of work undertaken in order to prepare specifically for these tests (Arvey et al., 1990)	I prepared a lot for the [specific test].*		

		Stakeholder Questions		
Topic	Topic Description	ALO and TACP Operators (Survey)	PTLs (Survey)	CFMs (Semistructured interviews)
Motivation	The amount of effort put forth during the testing process (Arvey et al., 1990)	I tried my best on every test.*		
Test administrator training	Training component included in the test administrator training delivered to the PTLs		I know how to assemble and operate the equipment for each test.	
			I know how to demonstrate and coach others on how to properly perform each test.	
			I know how to administer each test using the protocols.	
			I know how to score test performance using the score sheets.	
			I know how to use the standards to determine how well someone performed on each test.	
			I know the procedures to protect, store, and transfer the data.	
			I feel knowledgeable on how to properly administer the PF tests.	
			I feel adequately prepared to properly administer the PF tests.	
			I feel confident that I can properly administer the PF tests in the future.	

		Stakeholder Questions		
Topic	Topic Description	ALO and TACP Operators (Survey)	PTLs (Survey)	CFMs (Semistructured interviews)
Implementation considerations	Topics related to the implementation policy for the tests			What plans does your career field have for further evaluation or implementation of an updated physical test battery?
				What factors are most important in moving toward implementation?
				What barriers or obstacles are likely to interfere with moving toward implementation of an updated test battery (not necessarily the one currently proposed prototype)?

* Indicates that the item was not posed in the current administration; however, we would recommend posing it in future data collections.

Appendix C. TACP and ALO Evaluation Survey

Immediately after each test, TACPs and ALOs responded to four statements.

INSTRUCTIONS: On a 5-point scale, please evaluate your disagreement/agreement for each part of training. (*Circle one response for each item*)

	Strongly Disagree	Disagree	Neither Agree or Disagree	Agree	Strongly Agree
Grip Strength Test Component					
The grip strength test was administered to all operators in the same way.	SD	D	N A/D	A	SA
I am concerned that I was injured or could have been injured while completing the grip strength test.	SD	D	N A/D	A	SA
I know how I performed on the grip strength test relative to the required standard.	SD	D	N A/D	A	SA
Taking the grip strength test was frustrating.	SD	D	N A/D	A	SA
Medicine Ball Toss Test Component					
The medicine ball toss test was administered to all operators in the same way.	SD	D	N A/D	A	SA
I am concerned that I was injured or could have been injured while completing the medicine ball toss test.	SD	D	N A/D	A	SA
I know how I performed on the medicine ball toss test relative to the required standard.	SD	D	N A/D	A	SA
Taking the medicine ball toss test was frustrating.	SD	D	N A/D	A	SA
Three-Cone Drill Test Component					
The three-cone drill test was administered to all operators in the same way.	SD	D	N A/D	A	SA
I am concerned that I was injured or could have been injured while completing the three-cone drill test.	SD	D	N A/D	A	SA

	Strongly Disagree	Disagree	Neither Agree or Disagree	Agree	Strongly Agree
I know how I performed on the three-cone drill test relative to the required standard.	SD	D	N A/D	A	SA
Taking the three-cone drill test was frustrating.	SD	D	N A/D	A	SA
Trap Bar Deadlift Test Component The trap bar deadlift test was administered to all operators in the same way.	SD	D	N A/D	A	SA
I am concerned that I was injured or could have been injured while completing the trap bar deadlift test.	SD	D	N A/D	A	SA
I know how I performed on the trap bar deadlift test relative to the required standard.	SD	D	N A/D	A	SA
Taking the trap bar deadlift test was frustrating.	SD	D	N A/D	A	SA
Pull-Up Test Component The pull-up test was administered to all operators in the same way.	SD	D	N A/D	A	SA
I am concerned that I was injured or could have been injured while completing the pull-up test.	SD	D	N A/D	A	SA
I know how I performed on the pull-up test relative to the required standard.	SD	D	N A/D	A	SA
Taking the pull-up test was frustrating.	SD	D	N A/D	A	SA
Lunge Test Component The lunge test was administered to all operators in the same way.	SD	D	N A/D	A	SA
I am concerned that I was injured or could have been injured while completing the lunge test.	SD	D	N A/D	A	SA
I know how I performed on the lunge test relative to the required standard.	SD	D	N A/D	A	SA
Taking the lunge test was frustrating.	SD	D	N A/D	A	SA

66

	Strongly Disagree	Disagree	Neither Agree or Disagree	Agree	Strongly Agree
Extended Cross Knee Crunch Test Component The extended cross knee crunch test was administered to all operators in the same way.	SD	D	N A/D	A	SA
I am concerned that I was injured or could have been injured while completing the extended cross knee crunch test.	SD	D	N A/D	A	SA
I know how I performed on the extended cross knee crunch test relative to the required standard.	SD	D	N A/D	A	SA
Taking the extended cross knee crunch test was frustrating.	SD	D	N A/D	A	SA
Farmer's Carry Test Component The farmer's carry test was administered to all operators in the same way.	SD	D	N A/D	A	SA
I am concerned that I was injured or could have been injured while completing the farmer's carry test.	SD	D	N A/D	A	SA
I know how I performed on the farmer's carry test relative to the required standard.	SD	D	N A/D	A	SA
Taking the farmer's carry test was frustrating.	SD	D	N A/D	A	SA
Ergometer Row Test Component The ergometer row test was administered to all operators in the same way.	SD	D	N A/D	A	SA
I am concerned that I was injured or could have been injured while completing the ergometer row test.	SD	D	N A/D	A	SA
I know how I performed on the ergometer row test relative to the required standard.	SD	D	N A/D	A	SA
Taking the ergometer row test was frustrating.	SD	D	N A/D	A	SA

	Strongly Disagree	Disagree	Neither Agree or Disagree	Agree	Strongly Agree
1.5-Mile Run Test Component The 1.5-mile run was administered to all operators in the same way.	SD	D	N A/D	A	SA
I am concerned that I was injured or could have been injured while completing the 1.5-mile run.	SD	D	N A/D	A	SA
I know how I performed on the 1.5-mile run relative to the required standard.	SD	D	N A/D	A	SA
Taking the 1.5-mile run test was frustrating.	SD	D	N A/D	A	SA

After all tests are complete, TACPs and ALOs respond to the following questions.

INSTRUCTIONS: On a 5-point scale, please evaluate your disagreement/agreement for each part of training. (*Circle one response for each item*)

Item	Strongly Disagree	Disagree	Neither Agree or Disagree	Agree	Strongly Agree
This test battery clearly measures abilities required of a TACP.	SD	D	N A/D	A	SA
Doing well on this test battery means a person can perform the physical job duties of a TACP well.	SD	D	N A/D	A	SA
I could really show my physical abilities through this test battery.	SD	D	N A/D	A	SA
This test battery is a better measure of my operational capability than the PAST.	SD	D	N A/D	A	SA
Please tell us more about why you answered the previous question the way you did.					
I understood in advance what tests I would be taking.	SD	D	N A/D	A	SA
I have been provided with information about how my test results will be used.	SD	D	N A/D	A	SA
I know the consequences of not meeting test standards.	SD	D	N A/D	A	SA
I was able to ask questions about the tests before and during the tests.	SD	D	N A/D	A	SA
Test administrators answered questions in a straightforward manner.	SD	D	N A/D	A	SA
The test administrators treated me with respect during the testing process.	SD	D	N A/D	A	SA

I am satisfied with the feedback I received about my test results.	SD	D	N A/D	A	SA
Knowing how well I did on the PF tests will help me improve my job-related physical capability.	SD	D	N A/D	A	SA
Other TACPs will find their test results useful for improving their own job-related physical capability.	SD	D	N A/D	A	SA
The testing process I just completed would be fair for all TACPs regardless of rank, age, stature, gender, or race/ethnicity.	SD	D	N A/D	A	SA
It is important to me that the people who administer PF tests are other TACPs.	SD	D	N A/D	A	SA
I am confident that I am physically capable of doing my job as a TACP.	SD	D	N A/D	A	SA
I have concerns about this test battery.	SD	D	N A/D	A	SA

Please tell us more about your concerns.

Grip Strength _____

Med Ball Toss _____

Three Cone Drill _____

Trap Bar Deadlift _____

Lunges _____

Pull-Ups _____

Ext Cross Knee Crunch _____

Farmer's Carry _____

Erg Row, 1,000-m _____

Run, 1.5 Mile _____

Please tell us any changes that you would recommend to the testing.

69

BACKGROUND CHARACTERISTICS

Rank *(write in)*: _____

Component (Active/Guard/Reserve):
____Active
____Guard
____Reserve

Skill Level *(select one)*:
_____ 3 - Three Level
_____ 5 - Five Level
_____ 7 - Seven Level
_____ 9 - Nine Level
_____ CEM - CEM-1C200, 1C400, 1T200

Years in Career Field *(write in)*: _____

Height *(select one)*:
_____ < 5'0"
_____ 5'1" - 5'7"
_____ 5'8" - 5'11"
_____ 6'0" - 6'5"
_____ ≥ 6'6"

Weight *(select one)*:
_____ < 150 lbs
_____ 151 - 180 lbs
_____ 181 - 200 lbs
_____ 201 - 230 lbs
_____ > 230 lbs

Number of Deployments *(select one)*:
_____ 0
_____ 1 to 3
_____ 4 or 5
_____ 5+

Number of operational missions
(select one; estimate is fine):
_____ 0
_____ 1 - 25
_____ 26 – 50
_____ 51 or more

Appendix D. PTL Evaluation Survey

BACKGROUND: The purpose of training is to provide the necessary information to properly administer the battery of ten PF tests. To make sure training goals are met, we are examining different parts of training.

TEST/TRAINING FEEDBACK INSTRUCTIONS: On a 5-point scale, please evaluate your disagreement/agreement for each part of training. (*Circle one response for each item.*)

Item	Strongly Disagree	Disagree	Neither Agree or Disagree	Agree	Strongly Agree
1. I know how to assemble and operate the equipment for each test.	SD	D	N A/D	A	SA
2. I know how to demonstrate and coach others on how to properly perform each test.	SD	D	N A/D	A	SA
3. I know how to administer each test using the protocols.	SD	D	N A/D	A	SA
4. I know how to score test performance using the score sheets.	SD	D	N A/D	A	SA
5. I know how to use the standards to determine how well someone performed on each test.	SD	D	N A/D	A	SA
6. I know the procedures to protect, store, and transfer the data.	SD	D	N A/D	A	SA
7. I feel knowledgeable on how to properly administer the PF tests.	SD	D	N A/D	A	SA
8. I feel adequately prepared to properly administer the PF tests.	SD	D	N A/D	A	SA
9. I feel confident that I can properly administer the PF tests in the future.	SD	D	N A/D	A	SA

10. Please tell us a little more about what parts of training you feel may need to be changed and how:

_____
_____
_____
_____

ADDITIONAL QUESTIONS ABOUT THE TEST BATTERY

INSTRUCTIONS: On a 5-point scale, please evaluate your disagreement/agreement for each statement. (*Circle one response for each item*)

Item	Strongly Disagree	Disagree	Neither Agree or Disagree	Agree	Strongly Agree
I found the time available to inform TACPs about the exercise scoring criteria/expectations to be sufficient.	SD	D	N A/D	A	SA
Each test was administered to all operators in the same way.	SD	D	N A/D	A	SA
It appeared that taking this test battery was frustrating for TACPs.	SD	D	N A/D	A	SA
I am concerned that TACPs may have been injured or could have been injured while completing this test battery.	SD	D	N A/D	A	SA
Please elaborate below on any of your concerns for questions 11–14.					

Item	Strongly Disagree	Disagree	Neither Agree or Disagree	Agree	Strongly Agree
1. Operators could really show their physical abilities through this test battery.	SD	D	N A/D	A	SA
2. Doing well on this test battery means a person can perform the physical job duties of a TACP well.	SD	D	N A/D	A	SA
3. This test battery clearly measures abilities required of a TACP.	SD	D	N A/D	A	SA
4. This test battery is a better measure of operational capabilities than the PAST.	SD	D	N A/D	A	SA
5. Please tell us more about why you answered the previous question the way you did.					
6. TACPs will find their test results useful for improving their own job-related physical capability.	SD	D	N A/D	A	SA
7. The testing process would be fair for all TACPs regardless of rank, age, stature, gender, or race/ethnicity.	SD	D	N A/D	A	SA
8. It is important that the people who administer PF tests are other TACPs.	SD	D	N A/D	A	SA
9. I am confident that I am physically capable of doing my job as a TACP.	SD	D	N A/D	A	SA

10. I have concerns about this test battery.	SD	D	N A/D	A	SA

11. Please tell us more about your concerns.

Grip Strength _____

Med Ball Toss _____

Three Cone Drill _____

Trap Bar Deadlift _____

Lunges _____

Pull-Ups _____

Ext Cross Knee Crunch _____

Farmer's Carry _____

Erg Row, 1000-m _____

Run, 1.5 Mile _____

12. **Please tell us any changes that you would recommend to the testing.**

BACKGROUND CHARACTERISTICS

Rank *(write in)*: _____

Component (Active/Guard/Reserve):
____Active
____Guard
____Reserve

Skill Level (*select one*):
_____ 3 - Three Level
_____ 5 - Five Level
_____ 7 - Seven Level
_____ 9 - Nine Level
_____ CEM - CEM-1C200, 1C400, 1T200

Years in Career Field *(write in)*: ____

Height (*select one*):
____ < 5'0"
____ 5'1" - 5'7"
____ 5'8" - 5'11"
____ 6'0" - 6'5"
____ ≥ 6'6"

Weight (*select one*):
____ < 150 lbs
____ 151 - 180 lbs
____ 181 - 200 lbs
____ 201 - 230 lbs
____ > 230 lbs

Number of Deployments (*select one*):
____ 0
____ 1 to 3
____ 4 or 5
____ 5+

Number of operational missions
(select one; estimate is fine):
____ 0
____ 1 - 25
____ 26 – 50
____ 51 or more

Appendix E. Evaluation Framework

The following tables provide example issues and questions that could be addressed to provide information on how well tests and standards are implemented. Each table represents specific issues and questions for a specific level (e.g., individual level, unit level) and is organized using DOTMLPF-P (note that we replaced "doctrine" with "attitudinal" to more specifically address the unique needs of the communities studied). The frameworks for the individual, unit, career field, institutional, and societal level are provided in Tables E.1–E.5, respectively. For each topic and issue, we provide specific guidance on the type of data that can be used to evaluate the issue. For some issues, secondary data collection can be used, as data (e.g., medical records) may already exist in databases managed by the Air Force. For other issues, the Air Force may need to do primary data collection, such as interviews, focus groups, and surveys.

Table E.1. Evaluation Framework for Individual Level

Category	Topic	Issue	Metric	Method	Stakeholder Guidance
Organization	Readiness	How does individual readiness compare with readiness prior to introduction of the O-PTB?	Comparison with metrics prior to implementation of the O-PTB	Primary data collection	May need to develop new data-collection system
Training	Resources needed	How does the time needed to train for the O-PTB compare to the time needed prior to its implementation?	Comparison with metrics prior to implementation of the O-PTB	Primary data collection	May need to develop new data-collection system
Training	Injury	Does training for the O-PTB result in more injuries to test-takers compared to the PAST?	Comparison with metrics prior to implementation of the O-PTB	Secondary data collection	May need to develop new data-collection system if injuries are not well documented in medical records, for example
Materiel	Access to equipment	Do individuals have access to test equipment on which to train?	Review of equipment available at Air Force installations with BA	Primary data collection	May need to develop new data-collection system

Category	Topic	Issue	Metric	Method	Stakeholder Guidance
Leadership and Education	Guidance for performance improvement	Is the guidance provided for improving performance on the O-PTB sufficient for improving performance?	Perceptions of guidance; comparison of performance data pre- and post-provision of guidance	Primary data collection, secondary data collection	Surveys, focus groups, and interviews could be conducted among BA operators, PTLs, strength and conditioning coaches, and training instructors
Personnel	Entry	At what rates are individuals eligible to enter BA career fields relative to rates prior to introduction of the O-PTB?	Comparison with metrics prior to implementation of the O-PTB	Secondary data collection	
Personnel	Retention	At what rates are BA retained relative to rates prior to introduction of the O-PTB?	Comparison with metrics prior to implementation of the O-PTB	Secondary data collection	
Personnel	Promotion	At what rates are BA promoted relative to rates prior to introduction of the O-PTB?	Comparison with metrics prior to implementation of the O-PTB	Secondary data collection	
Personnel	Attrition	To what degree is failure to meet the O-PTB standards a reason for attrition?	Identify PF-related reasons for attrition	Primary data collection, secondary data collection	Surveys and interviews could be conducted among BA operators, CFMs
Personnel	Injury	Does the O-PTB result in more injuries to test-takers compared with prior tests?	Comparison with metrics prior to implementation of the O-PTB	Secondary data collection	
Personnel	Resources needed	How does the time needed to complete the O-PTB compare to the time needed prior to its implementation?	Comparison with metrics prior to implementation of the O-PTB	Primary data collection	May need to develop new data-collection system
Facilities	Access to facilities	Do individuals have access to facilities in which to train for the O-PTB?	Review of facilities available at Air Force installations with BA	Primary data collection	May need to develop new data-collection system
Policy	Policy	What policies regarding the O-PTB are in place for operators without a unit?	Identification of policies for operators without a unit	Policy review	

Category	Topic	Issue	Metric	Method	Stakeholder Guidance
Attitudinal	Test perceptions	What is the acceptability of the O-PTB among BA operators?	Identification of acceptability	Primary data collection	Interviews, focus groups, surveys could be conducted among BA operators, PTLs, CFMs
Attitudinal	Test perceptions	What are operator reactions to the O-PTB?	Identification of reactions to the O-PTB	Primary data collection	Interviews, focus groups, surveys could be conducted among BA operators, PTLs, CFMs
Attitudinal	Implementation perceptions	What are operator reactions to the process of implementing the O-PTB?	Identification of reactions to implementation process	Primary data collection	Interviews, focus groups, surveys could be conducted among BA operators, PTLs, CFMs, strength and conditioning coaches

77

Table E.2. Evaluation Framework for Unit Level

Category	Topic	Issue	Metric	Method	Stakeholder Guidance
Organization	Resources needed	How do the time and cost of conducting the PAST differ compared to time and cost of the O-PTB?	Comparison with metrics (e.g., number of PTLs required) prior to implementation of the O-PTB	Primary data collection	May need to develop new data-collection system
Organization	SMEs consulted	What SMEs were consulted during implementation and were they the most appropriate ones?	Identification of SMEs consulted and gaps in expertise	Primary data collection	May need to develop new data-collection system
Organization	Readiness	How does unit readiness compare with readiness prior to introduction of the O-PTB?	Comparison with metrics prior to implementation of the O-PTB	Primary data collection	May need to develop new data-collection system
Organization	Readiness	What are the perceived impacts of the O-PTB on unit readiness?	Identification of perceived impact	Primary data collection	Interviews, focus groups, surveys with BA operators, unit leadership, CFMs
Training	Training materials	What training materials have been created for PTLs?	Documentation of training materials	Materials review	
Training	Training effectiveness	How effective are training materials for providing PTLs with the necessary knowledge to administer the O-PTB?	Comparison of knowledge pre-training to post-training	Primary data collection	Surveys of PTLs
Training	Training effectiveness	If a train-the-trainer model is employed, how "drift" is there in test administration across PTLs?	Longitudinal comparison of training knowledge across PTLs	Primary data collection	Surveys of PTLs
Training	Resources needed	How do the time and cost of training for the PAST differ compared with the time and cost of training for the O-PTB?	Comparison with metrics prior to implementation of the O-PTB	Primary data collection	Interviews, focus groups with CFMs, PTLs, unit leadership
Materiel	Resources needed	What financial costs are associated with storing and maintaining equipment needed for testing?	Identification of costs for storage and maintenance	Primary data collection; secondary data collection	Interviews, focus groups with CFMs, PTLs, unit leadership

Category	Topic	Issue	Metric	Method	Stakeholder Guidance
Materiel	Resources needed	How often does testing equipment need to be recalibrated?	Identification of recalibration needs	Primary data collection	May need to develop new data-collection system
Leadership and Education	Guidance for performance improvement	Is guidance provided for improving performance on the O-PTB sufficient (for units with and without strength and conditioning coaches)?	Comparison of PF test performance before and after guidance Comparison across units with and without strength and conditioning coaches	Primary data collection; secondary data collection	Interviews with unit leadership, strength and conditioning coaches
Leadership and Education	Unit leadership support	Is unit leadership sufficiently and actively supporting implementation of the O-PTB?	Climate surveys, focus groups with operators	Primary data collection	Surveys and focus groups with BA operators in each unit
Leadership and Education	Unit leadership support	How have the O-PTB-related complaints been handled at the unit level?	Documentation of complaints, insight from unit leadership	Primary data collection; review of documentation	Interviews with unit leadership
Leadership and Education	Provision of scores to unit leadership	How are unit leaders provided with information about the performance of their operators? Is this information provided in a timely fashion? In a form that is easy to use?	Identify unit leader perceptions of score provision process	Primary data collection	Interviews with unit leadership
Facilities	Resources needed	Does introduction of the O-PTB require any unit-level facility changes?	Surveys of facility users, maintainers, and unit leadership	Primary data collection	Surveys of operators, maintainers of facilities, and unit leadership
Policy	Test data storage and transmission	What procedures and systems are in place for storing and transmitting test data during the implementation period and beyond? How easy are these systems to use? How secure?	Characterize policies surrounding test data storage and transmission	Policy review; primary data collection	Interviews with researchers and database administrators
Policy	Applicability across sites	Are instructions and policies for implementing tests sufficient for allowing implementation at a range of sites?	Characterize policies surrounding test implementation	Policy review, primary data collection	Interviews with BA unit leaders
Attitudinal	Cohesion and morale	What have been the effects of the O-PTB on unit cohesion and morale?	Unit climate	Primary data collection	Climate surveys, focus groups with operators and unit leaders

Table E.3. Evaluation Framework for Career Field Level

Category	Topic	Issue	Metric	Method	Stakeholder Guidance
Organization	Resources needed	How do the time and cost of conducting the PAST differ compared with the time and cost of the O-PTB?	Comparison with metrics prior to implementation of the O-PTB	Primary data collection	May need to develop new data-collection system
Organization	SMEs consulted	What SMEs were consulted during implementation and were they the most appropriate ones?	Documentation of SMEs consulted	Documentation	
Organization	Communication	Were communications about implementation of the O-PTB to operators within career fields effective in communicating necessary information?	Perception of communication efforts	Primary data collection	Interviews, focus groups with operators and CFMs
Organization	Communication	Did communications about implementation of the O-PTB to operators within a career field occur at an appropriate frequency?	Perception of communication efforts	Primary data collection	Interviews, focus groups with operators and CFMs
Organization	Readiness	How does readiness of BA within each career field compare with readiness prior to introduction of the O-PTB?	Comparison with pre–O-PTB metrics	Primary data collection	May need to develop new data-collection system
Organization	Readiness	What are the perceived impacts of the O-PTB on readiness of BA within each career field?	Identification of perceived impact	Primary data collection	Interviews, focus groups with CFMs
Training	Completion Rates	How have completion rates of career field training changed since introduction of the O-PTB?	Comparison with pre–O-PTB metrics	Secondary data collection	
Training	Resources needed	How do the time and cost of training for PF tests differ compared to the time and cost of training prior to implementation of the O-PTB?	Comparison with metrics prior to implementation of the O-PTB	Primary data collection	May need to develop new data-collection system
Leadership and Education	Career Development	What are the effects of the O-PTB on career development within each career field?	Career trajectory data, insight from CFMs	Primary data collection; secondary data collection	Interviews with CFMs

Category	Topic	Issue	Metric	Method	Stakeholder Guidance
Leadership and Education	Receiving Support	Are BA in each career field receiving the support and information they need to meet new PF standards?	Insight from operators and CFMs	Primary data collection	Interviews with CFMs, focus groups and surveys with BA operators
Personnel	Entry	Are enough people who meet the O-PTB standards entering a career field to allow for appropriate assignments?	Comparison of entrance rates pre- and post–O-PTB	Secondary data collection	
Personnel	Fairness	Is the O-PTB being administered and applied fairly to all BA in the career field?	Insights from operators and CFMs	Primary data collection	Surveys of operators, focus groups with CFMs
Personnel	Attrition	To what degree is failure to meet new PF standards a reason for attrition?	Identify PF-related reasons for attrition	Secondary data collection, interviews, surveys, focus groups	Surveys of operators; interviews, focus groups with CFMs
Policy	Grandfathering	How will policy be implemented for those already in training pipeline?	Review policy for relevant information	Policy review	
Policy	Duty versus Primary Air Force Specialty Code (AFSC)	How do policies account for duty AFSC compared with primary AFSC?	Review policy for relevant information	Policy review	
Policy	Phases	How do implementation policies account for implementation at different phases (e.g., recruitment, training)?	Review policy for relevant information	Policy review	
Policy	Operators Without Units	What policies are in place for operators without a unit?	Review policy for relevant information	Policy review	
Policy	Test administration and frequency	Do policies exist regarding test administration and frequency? Are they followed?	Review policy for relevant information	Policy review	
Policy	Individual tests	Do policies exist regarding individual tests within the test battery (e.g., allowing a current run time to be used or completing a new run within 72 hours of other PF tests)?	Review policy for relevant information	Policy review	

Category	Topic	Issue	Metric	Method	Stakeholder Guidance
Policy	Failure	What policies are in place regarding failure of the O-PTB? Are there different policies for the implementation period compared with after the implementation period?	Review policy for relevant information	Policy review	
Policy	Minimally Essential Times	Are the Minimally Essential Times set at the most appropriate level to allow for maximization of good selection/rejection and minimize missed opportunities and poor selection?	Review policy for relevant information	Policy review	
Policy	Use of test scores	What policies, if any, exist regarding the use of test scores for Enlisted Performance Reports, awards?	Review policy for relevant information	Policy review	
Attitudinal	Morale	What have been the effects of the introduction of the O-PTB on morale within career fields?	Insights from operators and CFMs	Primary data collection	Interviews and focus groups with CFMs, surveys with operators
Attitudinal	Test perceptions	How do leadership and BA operators within career fields differ in their perceptions of the O-PTB?	Insights from operators and CFMs	Primary data collection	Interviews and focus groups with CFMs, surveys with operators
Attitudinal	Test perceptions	What is the acceptability of the O-PTB within each career field?	Identification of acceptability	Primary data collection	Interviews and focus groups with CFMs, surveys with operators

Table E.4. Evaluation Framework for Institutional Level

Category	Topic	Issue	Metric	Method	Stakeholder Guidance
Organization	Resources needed	How do the time and cost of conducting PF tests differ compared with the time and cost prior to implementation of the O-PTB?	Comparison with metrics prior to implementation of the O-PTB	Primary data collection	May need to develop new data-collection system
Organization	SMEs consulted	What SMEs were consulted during implementation and were they the most appropriate ones?	Documentation of SMEs consulted	Documentation	
Organization	Communication	Were communications about implementation of the O-PTB to BA leadership effective in communicating necessary information?	Perception of communication efforts	Primary data collection	Interviews, focus groups with operators and CFMs
Organization	Communication	Did communications about implementation of the O-PTB to BA leadership occur at an appropriate frequency?	Perception of communication efforts	Primary data collection	Interviews, focus groups with operators and CFMs
Organization	Readiness	Does the new O-PTB improve readiness of Air Force BA?	Comparison with pre-O-PTB metrics	Primary data collection	May need to develop new data-collection system
Organization	Implementation period	Is the implementation period for the O-PTB the appropriate length?	Insight on implementation period	Primary data collection	Interviews with Air Force leaders
Organization	Implementation plan	Has the Air Force developed an implementation plan for the O-PTB?	Insight on implementation period	Primary data collection	Interviews with Air Force leaders
Personnel	Effects on other health behaviors	How does the implementation of new PF tests and training for such tests affect other operator health behaviors (e.g., nutrition needs)?	Insight on implications for other health behaviors	Primary data collection	Interviews with Exercise Science Unit staff
Personnel	Fairness	Are the tests fair and unbiased for different groups (e.g., gender, age, body mass)?	Insights from operators and CFMs	Primary data collection	Surveys of operators, interviews and focus groups with CFMs

83

Category	Topic	Issue	Metric	Method	Stakeholder Guidance
Facilities	Resources needed	Are there any facility changes required to facilitate the introduction of the O-PTB?	Surveys of facility users, maintainers, and unit leadership	Primary data collection	Surveys of operators, maintainers of facilities, and unit leadership
Policy	Implementation	Has an Air Force instruction been issued to address implementation of the O-PTB?	Review policy for relevant information	Policy review	
Policy	Testing frequency	How frequently is PF testing supposed to occur? Is this the most appropriate frequency? Does it actually occur at the frequency it is supposed to?	Review policy and test data for relevant information	Primary data collection; secondary data collection	
Policy	Implementation	Is the implementation plan being implemented according to the timeline?	Review policy and test data for relevant information	Policy review, secondary data collection	
Policy	Implementation	Is the implementation plan being executed within budget?	Review policy and budget information for relevant information	Policy review, secondary data collection	
Policy	Communication	Are internal and external communication plans about the O-PTB in place?	Review policy and communication plans for relevant information	Policy review, secondary data collection	
Policy	Communication	Are the communications plans being executed as stated?	Review policy and communication plans for relevant information	Policy review, secondary data collection	

Table E.5. Evaluation Framework for Societal Level

Evaluation Issue
Is the public aware of changes to operator test batteries?
How does changing the test battery affect public perceptions of the Air Force and of BA?
How does changing the test battery affect the role of specific groups within the Air Force (e.g., women, transgender individuals)?
How has the Air Force demonstrated an appreciation for diversity to the U.S. public?
Prior to being eligible for military service, how do opportunities for developing physical capabilities differ for men and women? That is, are men more likely to be physically capable of performing duties upon entry into the Air Force relative to women?
Prior to being eligible for military service, are men more likely to begin physically training for military PF tests relative to women? If so, why?
Should the Air Force provide information to potential recruits about the risks (e.g., of personal injury) of training for the test battery?
Is it appropriate to expect fitness improvements to occur on the same timeline for both male and female airmen?
How does changing the test battery affect perceptions among the general public of the tasks/jobs women are capable of performing?
How does changing the test battery affect partner nations' perceptions of the U.S. military, the Air Force, and BA?
How do any changes in the test battery affect long-term outcomes for BA operators when they transition to civilian status (e.g., in terms of physical and mental health conditions, healthcare costs, employability)?

Appendix F. TACP and ALO Open-Ended Comments

We reviewed all the open-ended comments, and two independent research associates identified meaningful themes for each question through an iterative process of classifying comments to ensure a comprehensive representation of participants' comments. Each comment was coded to the category it best reflected. This appendix provides the frequency counts of the categories and presents exemplar comments for each category. These comments are presented in the exact words of the respondents (i.e., they are unedited). Table F.1 provides comments on why operators felt the test battery was a better measure of operational capabilities compared to the PAST.

Table F.1. Please Tell Us More About Why You Answered the Previous Question the Way You Did (Previous Question: This Test Battery Is a Better Measure of My Operational Capability Than the PAST)

Coding Category	Examples	Counts
Good test	Extremely broad spectrum for all body types. The best overall fitness assessment I've participated in.	36
Other	It all depends on the situation. It's too soon to tell.	21
More comprehensive than PAST	It incorporates more aspects of fitness, especially strength. Trains more of a full body concept.	19
Better measure of physical abilities than PAST	PAST test was always minimum to operational requirements. This measures functional fitness. I think tells a lot more about a TACP's physical powers than push-ups, sit-ups, and a run.	18
Questioning validity of PAST or PAST component	The PAST is an outdated and irrelevant test for this career field.	11
Test is not reflective of job demands	Some exercises should be omitted or revised—the cross knee crunch should be revised—a lack of coordination/rhythm caused me to fail. I don't need rhythm to be a TACP.	9
Test is not comprehensive	Test should account for ruck march.	8
Cannot compare with PAST	I have never taken the PAST.	4
Different orientation/focus than PAST	More functional fitness oriented than the PAST. This test seems to be aimed more for TACP supporting unconventional forces.	4

Coding Category	Examples	Counts
Fitness not all that important	"Most" TACPs do not "need" to be fit. . . However, it helps to be more capable in the battlefield. Being a TACP is only half physical. The other half is ability to be a well-rounded operator (controlling air).	4
More challenging than PAST	More challenging.	3

The comments provided in Table F.2 are organized by frequency for the more negative concerns for this group of tests. The neutral and positive comments are provided at the bottom of the table.

Table F.2. Please Tell Us More About Your Concerns (About the Following Tests): Grip Strength, Medicine Ball Toss, Three Cone Drill, Trap Bar Deadlift, Lunges

Coding Category	Definition	Examples	Counts				
			Grip Strength	Medicine Ball Toss	Three-Cone Drill	Trap Bar Deadlift	Lunges
Importance of form/technique	Finds the exercise too focused on specific skills/technique.	I feel like my velocity is stagnated by the planting of my feet (for med ball toss).	1	21	4	0	0
More instruction/ practice needed	Would like more information on how to do the test and/or more practice to get acquainted with the test's technique.	Teaching good form is essential to the exercise (for med ball toss).	2	10	2	6	1
Risk of injury	Test presents a risk of injury.	(Trap bar deadlift) could be dangerous without proper training.	0	4	5	11	1
Testing environment	Mentions equipment, space, surface, and related environmental conditions such as weather.	The side throw is a little awkward, maybe use a smaller diameter ball (for medicine ball toss).	1	5	9	4	2
Purpose/utility unclear	Does not understand the purpose of the test or finds it useless.	I think pull-ups and deadlift can show enough grip strength.	14	3	1	0	0
Suggestion for improvement	Offers suggestion to improve the test.	Recommend measuring grip from more than one position (for grip strength).	4	5	0	4	2
Better test administration needed	Offers suggestion on how the test should be administered.	Need laser timer—human error can be very far off (for three-cone drill)	1	0	6	2	4

Coding Category	Definition	Examples	Counts				
			Grip Strength	Medicine Ball Toss	Three-Cone Drill	Trap Bar Deadlift	Lunges
Scoring criteria/ standards	Mentions scoring criteria and standards as an issue, and/or offers suggestions to change scoring criteria or standards.	The standards are pretty high for max high points (for lunges).	0	0	0	4	1
Unfair test	Sees different categories of operators as not having the same chances of passing.	Harder for taller/larger individuals to receive higher score (on three-cone drill).	0	2	2	1	0
Unclear how to improve	Comments on perceived difficulty in improving on the test.	Not sure how will improve on this (grip strength) other than technique.	3	0	1	0	0
Challenging test	Finds the test difficult and demanding.	Tough! (Lunges) smoked my ass!	0	0	0	0	8
Good test	Expresses satisfaction with the test.	(Trap bar is) great addition to the test.	3	2	3	6	3
Other	Other response that does not fit in other categories.	I got a low grip strength score but high pull-ups (for grip strength).	4	7	3	6	4

NOTE: Concerns with comments of ten or more in a category are highlighted in **bold** font.

The comments provided in Table F.3 are organized by frequency for the more negative concerns for this group of tests. The neutral and positive comments are provided at the bottom of the table.

Table F.3. Please Tell Us More About Your Concerns (About the Following Tests): Pull-Ups, Extended Cross Knee Crunch, Farmer's Carry, 1,000-Meter Ergometer Row, 1.5-Mile Run

Coding Category	Definition	Examples	Counts				
			Pull-Ups	Extended Cross Knee Crunch	Farmer's Carry	1,000-Meter Ergometer Row	1.5-Mile Run
Importance of form/technique	Finds the exercise too focused on specific skills/technique.	Awkward exercise, people were confused *(for extended cross knee crunch)*.	0	16	1	3	0
More instruction/ practice needed	Would like more information on how to do the test and/or more practice to get acquainted with the test's technique.	Most people have poor technique, we need a rowing clinic *(for 1,000-meter row)*.	2	15	0	2	0
Suggestion for improvement	Offers suggestion to improve the test.	Would prefer a 3-mile with body armor. 30-minute go/no go *(run)*.	0	5	5	2	7
Better test administration needed	Offers suggestion on how the test should be administered.	Make sure all testers adhere to the same form standards *(for pull-ups)*.	3	7	3	0	0
Testing environment	Mentions equipment, space, surface, weather conditions in relation to testing.	Smaller items to carry needed so they don't throw the operator off balance *(for farmer's carry)*.	0	1	9	0	0
Scoring criteria/ standards	Mentions scoring criteria and standards as an issue, and/or offers suggestions to change scoring criteria or standards.	Excessive for minimum standards *(for pull-ups)*.	5	3	1	0	0

Coding Category	Definition	Examples	Counts				
			Pull-Ups	Extended Cross Knee Crunch	Farmer's Carry	1,000-Meter Ergometer Row	1.5-Mile Run
Risk of injury	Test presents a risk of injury.	25-meter shuttle run dangerous on turns (during farmer's carry).	0	0	7	0	0
Purpose/utility unclear	Does not understand the purpose of the test or finds it useless.	No realistic function (for extended cross-knee crunch)	1	3	1	0	1
Unfair test	Sees different categories of operators as not having the same chances of passing.	N/A	0	0	0	0	0
Unclear how to improve	Offers suggestion on how the test should be administered.	N/A	0	0	0	0	0
Challenging test	Finds the test difficult and demanding.	Challenging. Loved it (1,000-meter row).	4	9	2	5	0
Good test	Expresses satisfaction with the test.	Good measure of cardio (1,000-meter row).	3	0	8	9	2
Other	Other response that does not fit in other categories.	Feel there could be another type of exercise for this (extended cross knee crunch).	5	7	3	3	8

NOTE: Concerns with comments of ten or more in a category are highlighted in **bold** font.

The comments provided in Table F.4 are organized by frequency and provide additional details on operators' recommended changes to testing.

Table F.4. Please Tell Us Any Changes That You Would Recommend to the Testing

Coding Category	Definition	Examples	Counts
Suggestion for improvement	Offers suggestion to improve the test.	Making the ruck mandatory because this is a crucial aspect of our career field. Add an extended endurance event on another day (12-mile run/ruck).	24
Scoring criteria/ standards	Mentions scoring criteria and standards as an issue, and/or offers suggestions to change scoring criteria or standards.	Smaller rep margins for cross knee and lunge. Based off of the science don't keep changing the scoring in the future unless it truly warrants it. It'll cheapen the results of those who took the test in the past.	12
Better test administration needed	Offers suggestion on how the test should be administered.	Take the human error out of timed events. Less metronome.	10
Other	Other response that does not fit in other categories.	I heard about the ruck and feel confident that if a member can pass this test he will be able to do a 12-km ruck. The ruck should not be combined with this test.	8
Testing environment	Mentions equipment, space, surface, weather conditions in relation to testing.	Guard needs more equipment to run the training.	6
Purpose/utility unclear	Does not understand the purpose of the test or finds it useless.	I am unsure why this (grip strength) is on the test. If your grip is good enough to pass the deadlift, then I think this is unnecessary.	6
Good test	Expresses satisfaction with the test.	Tests were performed in a timely manner. Sufficient rest was available between test components. I look forward to performing this test in the future.	6
More instruction/ practice needed	Would like more information on how to do the test and/or more practice to get acquainted with the test's technique.	Overall I just need (or this mix needs) more practice with specific tests.	5
Unfair test	Sees different categories of operators as not having the same chances of passing.	How much time would a member be given to build up to standards after a six-month or longer deployment if there are no available training facilities or operations tempo doesn't allow for specific training requirements?	4
Importance of form/technique	Finds the exercise too focused on specific skills/technique.	I would like to see the medicine ball techniques changed. Moving the feet would show true power.	3
Risk of injury	Test presents a risk of injury.	Having sports trainers would greatly help us to be able to train to minimize injury.	1

Appendix G. PTL Open-Ended Comments

We reviewed all the open-ended comments, and two independent research associates identified meaningful themes for each question through an iterative process of classifying comments to ensure a comprehensive representation of participants' comments. Each comment was coded to the category it best reflected. This appendix provides the frequency counts of the categories and presents exemplar comments for each category. These comments are presented in the exact words of the respondents (i.e., they are unedited). Table G.1 provides comments about what parts of training could be changed according to the PTLs.

Table G.1. Please Tell Us a Little More About What Parts of Training You Feel May Need to Be Changed and How

Coding Category	Definition	Examples	Counts
More instruction or practice needed	Requests more information on how to do the test and/or more practice to get acquainted with the test's technique.	Add videos to show form. More clarification on medicine ball toss—can we slide feet when throwing?	18
Change scoring or standards	Mentions scoring criteria and standards as an issue, and/or offers suggestions to change scoring criteria or standards.	More realistic pull-up scores/goals. Hardly anyone can max out at 30 pull-ups. I would allow test-taking members an opportunity to re-fire events.	16
Other	Other response that does not fit in other categories.	Will need to administer the test a few times before answering. As long as units have the same gear the test will stay valid. If units start ordering different gear, then the results might be slightly different.	9
Change one exercise	Mentions changing an exercise.	Replace pull-ups with weighted pull-ups. Ball toss: allow member to fall forward on the back toss.	9
No change required	Positive feedback, no change needed.	Instructors did a great job demonstrating tests.	8
Better test administration needed	Offers suggestion on how the test should be administered.	Only concern is consistency in testing between evaluators when evaluators or testers are from same unit. There should be a start and end time for each event.	7

Table G.2 provides additional information on why PTLs thought the test battery would be a better measure of operational capabilities compared to the PAST.

Table G.2. Please Tell Us More About Why You Answered the Previous Question (#18) the Way You Did: "This Test Battery Is a Better Measure of Operational Capabilities Than the PAST"

Coding Category	Examples	Counts
Other	Allow members to run on the ruck.	15
Test is better than PAST	This test far exceeds the PAST test. We need this test implemented now! Battlefield physiology requires more than just calisthenics and cardio based performance; this test encompasses the strength, endurance, and stamina needed to perform the job at the required physical level.	14
Good test/good measure of aptitudes	I truly believe that this test will improve the TACP. It will trim the fat of the career field. Clearly shows physical aptitude and overall PF.	13
Test is relevant to job	I noticed that the test highlighted areas I'm strong in the field. Buddy carries and ruck translated into a good farmer's carry and deadlift score. I could see the test correlate to the on-the-job performance.	10
Test is more comprehensive than PAST	This is a more rounded test to assess operator abilities.	9
Purpose/utility unclear	It's hard to explain "why" these exercises are relevant to the job.	8
Test better reflects job demands than PAST	The test definitely felt more geared toward what we actually do while on mission. The movements in the test are more relevant to job requirements.	7
Problem with standards	I do feel that the pull-ups are a little high.	5
More instruction/ practice needed	Some of the exercises you need to practice at it for a longer amount of time to get a rhythm and understand how to properly do it, like the medicine ball toss for balance and the farmer's carry (need to practice technique for better results).	3
Not a good measure of fitness/aptitudes	I feel some of the tests depend on technique as well. They may be strong, but if they have bad form, then it'll affect their score.	3

Coding Category	Examples	Counts
Not all tests required	Depends on the Joint Terminal Attack Controller's position if he could be required to do the whole test.	3
Indicates uncertainty	At this stage it would be difficult to say if this test can measure an operator's skill and fitness. But it's a step in the right direction.	3

The comments provided in Table G.3 are organized by frequency for the more negative concerns for this group of tests. The neutral and positive comments are provided at the bottom of the table. Concerns with comments of seven or more in a category are highlighted in bold.

Table G.3. Please Tell Us More About Your Concerns About the Following Tests: Grip Strength, Medicine Ball Toss, Three Cone Drill, Trap Bar Deadlift, Lunges

Coding Category	Definition	Examples	Counts				
			Grip Strength	Medicine Ball Toss	Three-Cone Drill	Trap Bar Deadlift	Lunges
Testing environment	Mentions equipment, space, surface, weather conditions in relation to testing.	Standardize the surface for the test *(for three-cone drill)*.	0	1	**8**	4	0
Risk of injury	Test presents a risk of injury.	Sudden stops jarring the knees *(for three-cone drill)*.	0	1	2	5	2
Importance of form/technique	Finds the exercise too focused on specific skills/technique.	*(For medicine ball toss)*, I think feet should be able to come off the ground. Allow for more explosive movement, needs a standard surface.	1	**7**	0	1	0
Purpose/utility unclear	Does not understand the purpose of the test or finds it useless.	*(Grip strength is)* useless! Due to other events that involve grip!	5	1	0	0	0
Scoring criteria/ standards	Mentions scoring criteria and standards as an issue, and/or offers suggestions to change scoring criteria or standards.	Simplify the scoring so there's no possibility of error *(for trap bar deadlift)*.	0	0	0	5	1

Coding Category	Definition	Examples	Counts				
			Grip Strength	Medicine Ball Toss	Three-Cone Drill	Trap Bar Deadlift	Lunges
Better test administration needed	Offers suggestion on how the test should be administered.	Calibrating could cause issues *(for grip strength)*.	2	0	3	1	0
Suggestion for improvement	Offers suggestion to improve the test.	Would prefer flat bar *(for trap bar deadlift)*.	1	1	0	4	0
More instruction/ practice needed	Would like more information on how to do the test and/or more practice to get acquainted with the test's technique.	Good coaching needs to occur prior to testing *(for trap bar deadlift)*.	0	1	0	4	0
Challenging test	Finds the test difficult and demanding.	Hard *(lunges)*.	0	0	0	0	1
Good test	Expresses satisfaction with the test.	Good event *(three-cone drill)*.	0	0	1	0	0
Other	Other response that does not fit in other categories but with too few mentions to have its own category.	Need to improve power personally *(for med ball toss)*.	5	4	3	2	2

NOTE: Concerns with comments of ten or more in a category are highlighted in bold font.

The comments provided in Table G.4 are organized by frequency for the more negative concerns for this group of tests. The neutral and positive comments are provided at the bottom of the table. There were no concerns with comments of seven or more for any test in a specific category. Table G.5 provides additional comments on recommended changes to make to training.

Table G.4. Please Tell Us More About Your Concerns About the Following Tests: Pull-Ups, Extended Cross Knee Crunch, Farmer's Carry, 1,000-Meter Ergometer Row, Run, 1.5-Mile

Coding Category	Definition	Examples	Counts				
			Pull-Ups	Extended Cross Knee Crunch	Farmer's Carry	1,000-Meter Erg Row	1.5-Mile Run
Scoring criteria/ standards	Mentions scoring criteria and standards as an issue, and/or offers suggestions to change scoring criteria or standards.	I thought the ten-point standard and passing standard were a little high (for pull-ups).	3	5	3	0	0
Importance of form/technique	Finds the exercise too focused on specific skills/technique.	Most people row inefficiently (for 1,000-meter row).	0	6	0	1	0
Testing environment	Mentions equipment, space, surface, weather conditions in relation to testing.	100-yard (farmer's) carry is difficult in most facilities to administer.	1	1	3	0	0
More instruction/ practice needed	Would like more information on how to do the test and/or more practice to get acquainted with the test's technique.	Better explanation of the thumb location under arm (for extended cross knee crunch).	0	5	0	0	0
Risk of injury	Test presents a risk of injury.	I'm concerned about possible injury (for farmer's carry).	0	0	4	0	0
Purpose/utility unclear	Does not understand the purpose of the test or finds it useless.	Not sure this (extended cross knee crunch) is the best measure of core abilities.	0	2	1	0	0
Better test administration needed	Offers suggestion on how the test should be administered.	Standard for width of hands should be upheld (for pull-ups).	1	1	1	0	0

Coding Category	Definition	Examples	Counts				
			Pull-Ups	Extended Cross Knee Crunch	Farmer's Carry	1,000-Meter Erg Row	1.5-Mile Run
Suggestion for improvement	Offers suggestion to improve the test.	Make (1.5-mile run) longer 3-mile run.	0	0	1	0	2
Unfair test	Sees different categories of operators as not having the same chances of passing.	N/A	0	0	0	0	0
Unclear how to improve	Offers suggestion on how the test should be administered.	N/A	0	0	0	0	0
Challenging test	Finds the test difficult and demanding.	(1,000-meter row) will get a lot of people.	0	2	1	2	1
Good test	Expresses satisfaction with the test.	Cool (pull-ups).	1	1	0	0	0
Other	Other response that does not fit in other categories but with too few mentions to have its own category.	Prefer (farmer's carry) to the shuttle.	2	3	5	0	1

Table G.5. Please Tell Us Any Changes That You Would Recommend to the Testing

Coding Category	Definition	Examples	Counts
Suggestion for improvement	Offers suggestion to improve the test.	Make it a continuous circuit with different exercise such as buddy drag, low crawl, and so on. Have a load out for the ruck vs. weight. More events possibly and should test be taken in combat uniform? Could the medicine ball toss be replaced with a power clean? Less awkward movement that doesn't require as much "skill." It might help TACPs develop proper lifting techniques as well. Many people already lift without proper coaching.	9
Scoring criteria/ standards	Mentions scoring criteria and standards as an issue, and/or offers suggestions to change scoring criteria or standards.	Pass/fail standard for some or all of the events. Study rank-based standards NCO [Non-Commissioned Officer] / SNCO [Senior Non-Commissioned Officer] CGO [Company Grade Officer] / FGO [Field Grade Officer]. In reality, the jobs are different at higher ranks and army echelons. Replace Air Force PT [physical training] with this test for BA.	8
Test administration	Offers suggestion on how the test should be administered.	Honestly I think that non-TACP persons that understand the importance both physically and career-wise could be a better test admin because they are less likely to be lax on scoring and form.	5
Unfair test	Sees different categories of operators as not having the same chances of passing.	Not sure how this will be gauged against current TACP with real battlefield injuries, i.e., slipped disk from deployment.	3
More instruction/practice needed	Would like more information on how to do the test and/or more practice to get acquainted with the test's technique.	More clarity on the scoring standards and form required for each exercise.	2
Testing environment	Mentions equipment, space, surface, weather conditions in relation to testing.	I would suggest a sled or prowler for the farmer's carry.	1
Risk of injury	Test presents a risk of injury.	Certified strength/conditioning coaches to help administer the TBDL, as improper form/coaching could lead to serious injury.	1
Good test	Expresses satisfaction with the test.	This test is better than the PAST test. Lunges, crunches, and row are great things to test people on. It shows how much endurance they have.	4
Other	Response that does not fit in other categories but with too few mentions to have its own category.	Would recommend if individuals fail to pass and fail to pass reassessments not be removed from career field, but become nondeployable at brigade level or Duties Not to Include Controlling.	6

98

References

AcqNotes, "JCIDS Process: JCIDS Manual of Operations," updated July 12, 2017. As of March 15, 2018:
http://acqnotes.com/acqnote/acquisitions/jcids-manual-operations

Air Force Personnel Center, *Air Force Enlisted Classification Directory: The Official Guide to the Air Force Enlisted Classification Codes*, October 31, 2016.

Air Force Personnel Center, *Air Force Officer Classification Directory: The Official Guide to the Air Force Officer Classification Codes*, October 31, 2017.

Alliger, G. M., and E. A. Janak, "Kirkpatrick's Levels of Training Criteria: Thirty Years Later," *Personnel Psychology*, Vol. 42, No. 2, June 1989, pp. 331–342.

Alliger, G. M., S. I. Tannenbaum, W. Bennett, H. Traver, and A. Shotland, "A Meta-Analysis of the Relations Among Training Criteria," *Personnel Psychology*, Vol. 50, No. 2, 1997, pp. 341–358.

Arvey, R. D., W. Strickland, G. Drauden, and C. Martin, "Motivational Components of Test Taking," *Personnel Psychology*, Vol. 43, No. 4, December 1990, pp. 695–716.

Bauer, T. N., J. McCarthy, N. Anderson, D. M. Truxillo, and J. F. Salgado, "What We Know About Applicant Reactions on Attitudes And Behavior: Research Summary and Best Practices," Society for Industrial and Organizational Psychology White Paper Series, 2012.

Bauer, T. N., D. M. Truxillo, R. J. Sanchez, J. M. Craig, P. Ferrara, and M. A. Campion, "Applicant Reactions to Selection: Development of The Selection Procedural Justice Scale (SPJS)," *Personnel Psychology*, Vol. 54, No. 2, June 2001, pp. 387–419.

Brown, K. G., "An Examination of the Structure and Nomological Network of Trainee Reactions: A Closer Look at 'Smile Sheets,'" *Journal of Applied Psychology*, Vol. 90, No. 5, 2005, pp. 991–1001.

Chairman of the Joint Chiefs of Staff, "Women in the Service Implementation Plan," memorandum, Washington, D.C., January 9, 2013.

Chan, D., N. Schmitt, J. M. Sacco, and R. P. DeShon, "Understanding Pretest and Posttest Reactions to Cognitive Ability and Personality Tests," *Journal of Applied Psychology*, Vol. 83, No. 3, 1998, pp. 471–485.

Colquitt, Jason A., and Kate P. Zipay, "Justice, Fairness, and Employee Reactions," *Annual Review of Organizational Psychology and Organizational Behavior*, Vol. 2, No. 1, 2015, pp. 75–99.

Defense Acquisition University, "DOTmLPF-P Analysis," webpage, October 30, 2017. As of December 29, 2017:
https://www.dau.mil/acquipedia/Pages/ArticleDetails.aspx?aid=d11b6afa-a16e-43cc-b3bb-ff8c9eb3e6f2

Derous, E., M. P. Born, and K. D. Witte, "How Applicants Want and Expect to Be Treated: Applicants' Selection Treatment Beliefs and the Development of the Social Process Questionnaire on Selection," *International Journal of Selection and Assessment*, Vol. 12, No. 1–2, March 2004, pp. 99–119.

Derous, E., and B. Schreurs, "Modeling the Structure of Applicant Reactions: An Empirical Study Within the Belgian Military," *Military Psychology*, Vol. 21, No. 1, 2009, pp. 40–61.

Fiske, S. T., and S. E. Taylor, *Social Cognition*, New York: McGraw-Hill, 1991.

Gilliland, S. W., "The Perceived Fairness of Selection Systems: An Organizational Justice Perspective," *Academy of Management Review*, Vol. 18, No. 4, 1993, pp. 694–734.

Greenberg, J., "Organizational Justice: Yesterday, Today, and Tomorrow," *Journal of Management*, Vol. 16, No. 2, 1990, pp. 399–432.

Hausknecht, J. P., D. V. Day, and S. C. Thomas, "Candidate Reactions to Selection Procedures: An Updated Model and Meta-Analysis," *Personnel Psychology*, Vol. 57, No. 3, September 16, 2004, pp. 639–683.

Hülsheger, Ute R., and Neil Anderson, "Applicant Perspectives in Selection: Going Beyond Preference Reactions," *International Journal of Selection and Assessment*, Vol. 17, No. 4, 2009, pp. 335–345.

Kirkpatrick, D., "Revisiting Kirkpatrick's Four-Level-Model," *Training and Development,* No. 1, 1996, pp. 54–57.

Kirkpatrick, D. L., "Evaluation of Training," in R. L. Craig, ed., *Training and Development Handbook: A Guide to Human Resource Development*, 2nd ed., New York: McGraw-Hill, 1976.

Mathieu, J. E., S. L. Tannenbaum, and E. Salas, "Influences of Individual and Situational Characteristics on Measures of Training Effectiveness," *Academy of Management Journal*, Vol. 35, No. 4, 1992, pp. 828–847.

McCarthy, Julie, Chad H. Van Iddekinge, Filip Lievens, Mei-Chuan Kung, Evan F. Sinar, and Michael Campion, "Do Candidate Reactions Relate to Job Performance or Affect Criterion-Related Validity? A Multistudy Investigation of Relations Among Reactions, Selection Test

Scores, and Job Performance," *Journal of Applied Psychology*, Vol. 98, No. 5, 2013, pp. 701–719.

Morgan, R.B., and W. J. Casper, "Examining the Factor Structure of Participant Reactions to Training: A Multidimensional Approach," *Human Resource Development Quarterly*, Vol. 11, No. 3, 2000, pp. 301–317.

Patel, L., *ASTD State of the Industry Report 2010*, Alexandria, Va.: American Society for Training and Development, 2010.

Phillips, J. J., *Handbook of Training Evaluation and Measurement Methods*, Houston, Tex.: Gulf, 1997.

Ployhart, R. E., and C. M. Harold, "The Applicant Attribution-Reaction Theory (AART): An Integrative Theory of Applicant Attributional Processing," *International Journal of Selection and Assessment*, Vol. 12, No. 1–2, 2004, pp. 84–98.

Robson, Sean, Maria C. Lytell, Anthony Atler, Jason H. Campbell, and Carra S. Sims, *Physical Task Simulations: Performance Measures for the Validation of Physical Tests and Standards for Battlefield Airmen*, Santa Monica, Calif.: RAND Corporation, RR-1595-AF, forthcoming.

Ryan, A. M., and R. E. Ployhart, "Applicants' Perceptions of Selection Procedures and Decisions: A Critical Review and Agenda for the Future," *Journal of Management*, Vol. 26, No. 3, 2000, pp. 565–606.

Salas, Eduardo, Scott I. Tannenbaum, Kurt Kraiger, and Kimberly A. Smith-Jentsch, "The Science of Training and Development in Organizations What Matters in Practice," *Psychological Science*, Vol. 13, No. 2, 2012, pp. 74–101.

Schaefer, Agnes Gereben, Jennie W. Wenger, Jennifer Kavanagh, Gillian S. Oak, Jonathan P. Wong, Thomas E. Trail, and Todd Nichols, *Implications Of Integrating Women into the Marine Corps Infantry*, Santa Monica, Calif.: RAND Corporation, RR-1103-USMC, 2015. As of April 13, 2018:
https://www.rand.org/pubs/research_reports/RR1103.html

Schmit, M. J., and A. M. Ryan, "Test-Taking Dispositions: A Missing Link?" *Journal of Applied Psychology*, Vol. 77, No. 5, 1992, pp. 629–637.

Schuler, H., "Social Validity of Selection Situations: A Concept and Some Empirical Results," in H. Schuler, J. L. Farr, and M. Smith M., eds., *Personnel Selection and Assessment: Individual and Organizational Perspectives*, Hillsdale, N.J.: Lawrence Erlbaum Associates, Inc., 1993, pp. 11–26.

Sitzmann, T., K. Brown, W. J. Casper, K. Ely, and R. D. Zimmerman, "A Review and Meta-Analysis of the Nomological Network of Trainee Reactions," *Journal of Applied Psychology*, Vol. 93, No. 2, 2008, pp. 280–295.

Tracey, J. B., T. R. Hinkin, S. Tannenbaum, and J. E. Mathieu, "The Influence of Individual Characteristics and the Work Environment on Varying Levels of Training Outcomes," *Human Resource Development Quarterly*, Vol. 12, No. 1, 2001, pp. 5–23.

Truxillo, D. M., T. Bodner, M. Bertolino, T. N. Bauer, and C. Yonce, "Effects of Explanations on Applicant Reactions: A Meta-Analytic Review," *International Journal of Selection and Assessment*, Vol. 17, No. 4, 2009, pp. 346–361.

U.S. Air Force, "Air Liaison Officer," undated. As of October 23, 2017:
https://www.airforce.com/careers/detail/air-liaison-officer

Warr, Peter, and David Bunce, "Trainee Characteristics and the Outcomes of Open Learning," *Personnel Psychology*, Vol. 48, No. 2, 1995, pp. 347–375.